FOOD PACKAGING TECHNIQUES

The Author

Dr. Z. A. Fakunle served as professor of Food Science and Nutrition at University of Nigeria, Nsukka. Currently, Dr. Fakunle Nutritional Consultant. Dr. Fakunle has to his credit about 100 research papers, 12 invited papers in National and International conferences, 24 book chapters and numbers of training manuals, reports and review.

FOOD PACKAGING TECHNIQUES

Dr. Z. A. Fakunle

2025

Agri Horti Press

New Delhi-110 084

ISBN 978-93-87642-72-0 (Hardbound)

Published By

Agri Horti Press

New Delhi

E-mail: agrihortipress@gmail.com

Digitally Printed at : Replika Press Pvt. Ltd.

PREFACE

Food packaging is packaging for food. A package provides protection, tampering resistance, and special physical, chemical, or biological needs. It may bear a nutrition facts label and other information about food being offered for sale. A choice of packaging machinery includes technical capabilities, labor requirements, worker safety, maintainability, serviceability, reliability, ability to integrate into the packaging line, capital cost, floorspace, flexibility, energy usage, quality of outgoing packages, qualifications, throughput, efficiency, productivity, ergonomics, etc. It is critical to maintain Food safety during processing, packaging, storage, logistics, sale, and use. Conformance to applicable regulations is mandatory. Some are country specific such as the US Food and Drug Administration and the US Department of Agriculture; others are regional such as the European Food Safety Authority. Certification programs such as the Global Food Safety Initiative are sometimes used. Food packaging considerations may include: use of Hazard analysis and critical control points, Verification and validation protocols, Good manufacturing practices, use of an effective Quality management system, Track and trace systems, and requirements for label content. Special Food contact materials are used when the package is in direct contact with the food product. Depending on the packaging operation and the food, packaging machinery often needs specified daily wash-down and cleaning procedures. Food Packaging plays a vital role in preserving food throughout the distribution chain. Without packaging, the processing of food can become compromised as it is contained by direct contact with physical, chemical and biological contaminants. The development of novel food packaging has not only increased the shelf life of foods, but also their safety and quality- therefore bringing convenience to consumers. Directly related, and interlinked, with food packaging is the concept of shelf life- the length of time that food, beverages, pharmaceutical drugs, chemicals, and many other perishable items are given before they are considered unsuitable for sale, use, or consumption. The present book contains consumer perception of food biotechnology, safe handling of food, food storage, packaging of fresh and processed meat, modern packaging for fresh poultry, the expanded roles of food and beverage packaging, dairy process biotechnology and products packaging. Important reference have also been appended at the end of book. I would look forward for comments and suggestions so that could be prepared better.

Z. A. Fakunle

Contents

1

Packaging and Labelling: An Introduction

Concept of Packaging and Labelling

Packaging is the science, art and technology of enclosing or protecting products for distribution, storage, sale, and use. Packaging also refers to the *process* of design, evaluation, and production of packages. Packaging can be described as a *coordinated system* of preparing goods for transport, warehousing, logistics, sale, and end use. Packaging contains, protects, preserves, transports, informs, and sells. In many countries it is fully integrated into government, business, institutional, industrial, and personal use. I LIKE PIE Package labelling or labelling is any written, electronic, or graphic communications on the packaging or on a separate but associated label.

The Purposes of Packaging and Package Label

Packaging and package labelling have several objectives:

- Physical protection - The objects enclosed in the package may require protection from, among other things, shock, vibration, compression, temperature, etc.
- Barrier protection - A barrier from oxygen, water vapour, dust, etc., is often required. Permeation is a critical factor in design. Some packages contain desiccants or Oxygen absorbers to help extend shelf life. Modified atmospheres or controlled atmospheres are also maintained in some food packages. Keeping the contents clean, fresh, sterile and safe for the intended shelf life is a primary function.
- Containment or agglomeration - Small objects are typically grouped together in one package for reasons

of efficiency. For example, a single box of 1000 pencils requires less physical handling than 1000 single pencils. Liquids, powders, and granular materials need containment.

- Information transmission - Packages and labels communicate how to use, transport, recycle, or dispose of the package or product. With pharmaceuticals, food, medical, and chemical products, some types of information are required by governments. Some packages and labels also are used for track and trace purposes.
- Marketing - The packaging and labels can be used by marketers to encourage potential buyers to purchase the product. Package graphic design and physical design have been important and constantly evolving phenomenon for several decades. Marketing communications and graphic design are applied to the surface of the package and the point of sale display.
- Security - Packaging can play an important role in reducing the security risks of shipment. Packages can be made with improved tamper resistance to deter tampering and also can have tamper-evident features to help indicate tampering. Packages can be engineered to help reduce the risks of package pilferage: Some package constructions are more resistant to pilferage and some have pilfer indicating seals.

  Packages may include authentication seals and use security printing to help indicate that the package and contents are not counterfeit. Packages also can include anti-theft devices, such as dye-packs, RFID tags, or electronic article surveillance tags that can be activated or detected by devices at exit points and require specialized tools to deactivate. Using packaging in this way is a means of loss prevention.
- Convenience - Packages can have features that add convenience in distribution, handling, stacking, display, sale, opening, reclosing, use, dispensing, and reuse.

- Portion control - Single serving or single dosage packaging has a precise amount of contents to control usage. Bulk commodities can be divided into packages that are a more suitable size for individual households. It is also aids the control of inventory: selling sealed one-liter-bottles of milk, rather than having people bring their own bottles to fill themselves! i like pie

Packaging Types

Packaging may be looked at as being of several different types. For example a transport package or distribution package can be the shipping container used to ship, store, and handle the product or inner packages. Some identify a consumer package as one which is directed toward a consumer or household. Packaging may be described in relation to the type of product being packaged: medical device packaging, bulk chemical packaging, over-the-counter drug packaging, retail food packaging, military materiel packaging, pharmaceutical packaging, etc.

It is sometimes convenient to categorize packages by layer or function: "primary", "secondary", etc.:

- Primary packaging is the material that first envelops the product and holds it. This usually is the smallest unit of distribution or use and is the package which is in direct contact with the contents.
- Secondary packaging is outside the primary packaging, perhaps used to group primary packages together.
- Tertiary packaging is used for bulk handling, warehouse storage and transport shipping. The most common form is a palletized unit load that packs tightly into containers.

These broad categories can be somewhat arbitrary. For example, depending on the use, a shrink wrap can be primary packaging when applied directly to the product, secondary packaging when combining smaller packages, and tertiary packaging on some distribution packs.

Symbols Used on Packages and Labels

Many types of symbols for package labelling are nationally

and internationally standardized. For consumer packaging, symbols exist for product certifications, trademarks, proof of purchase, etc. Some requirements and symbols exist to communicate aspects of consumer use and safety. Examples of environmental and recycling symbols include: Recycling symbol, Resin identification code, and Green Dot.

Bar codes, Universal Product Codes, and RFID labels are common to allow automated information management in logistics and retailing. Country of Origin Labelling is often used.

Shipping Container Labelling

Technologies related to shipping containers are identification codes, bar codes, and electronic data interchange. These three core technologies serve to enable the business functions in the process of shipping containers throughout the distribution channel. Each has an essential function: identification codes either relate product information or serve as keys to other data, bar codes allow for the automated input of identification codes and other data, and EDI moves data between trading partners within the distribution channel.

Elements of these core technologies include UPC and EAN item identification codes, the SCC-14, the SSCC-18, Interleaved 2-of-5 and UCC/EAN-128 bar code symbologies, and ANSI ASC X12 and UN/EDIFACT EDI standards. Small parcel carriers often have their own formats. For example, United Parcel Service has a MaxiCode 2-D code for parcel tracking.

RFID labels for shipping containers are also increasing in usage. A Wal-Mart division, Sam's Club, has also moved in this direction and is putting pressure on its suppliers for compliance. Shipments of hazardous materials or dangerous goods have special information and symbols as required by UN, country, and specific carrier requirements. Two examples are below:

With transport packages, standardised symbols are also used to communicate handling needs. Some common ones are shown below while others are listed in ASTM D5445 "Standard Practice for Pictorial Markings for Handling of Goods" and ISO 780 "Pictorial marking for handling of goods".

Fig. Fragile

Fig. Do not use hand hooks

Fig. This way up

Fig. Keep away from sunlight

Fig. Keep away from water

Fig. Centre of gravity

Fig. Clamp as indicated

Fig. Do not clamp as indicated

Package Development Considerations

Package design and development are often thought of as

an integral part of the new product development process. Alternatively, development of a package can be a separate process, but must be linked closely with the product to be packaged. Package design starts with the identification of all the requirements: structural design, marketing, shelf life, quality assurance, logistics, legal, regulatory, graphic design, end-use, environmental, etc. The design criteria, performance, completion time targets, resources, and cost constraints need to be established and agreed upon.

An example of how package design is affected by other factors is the relationship to logistics. When the distribution system includes individual shipments by a small parcel carrier, the sortation, handling, and mixed stacking make severe demands on the strength and protective ability of the transport package. If the logistics system consists of uniform palletized unit loads, the structural design of the package can be designed to those specific needs: vertical stacking, perhaps for a longer time frame. A package designed for one mode of shipment may not be suited for another.

With some types of products, the design process involves detailed regulatory requirements for the package. For example with packaging foods, any package components that may contact the food are food contact materials. Toxicologists and food scientists need to verify that the packaging materials are allowed by applicable regulations. Packaging engineers need to verify that the completed package will keep the product safe for its intended shelf life with normal usage. Packaging processes, labelling, distribution, and sale need to be validated to comply with regulations and have the well being of the consumer in mind.

Sometimes the objectives of package development seem contradictory. For example, regulations for an over-the-counter drug might require the package to be tamper-evident and child resistant: These intentionally make the package difficult to open.. The intended consumer, however, might be handicapped or elderly and be unable to readily open the package. Meeting all goals is a challenge.

Package design may take place within a company or with various degrees of external packaging engineering: independent contractors, consultants, vendor evaluations, independent laboratories, contract packagers, total outsourcing, etc. Some sort of formal Project planning and Project management methodology is required for all but the simplest package design and development programs. An effective quality management system and Verification and Validation protocols are mandatory for some types of packaging and recommended for all.

Package development involves considerations for sustainability, environmental responsibility, and applicable environmental and recycling regulations. It may involve a life cycle assessment which considers the material and energy inputs and outputs to the package, the packaged product, the packaging process, the logistics system, waste management, etc. It is necessary to know the relevant regulatory requirements for point of manufacture, sale, and use.

The traditional "three R's" of reduce, reuse, and recycle are part of a waste hierarchy which may be considered in product and package development:

- Prevention – Waste prevention is a primary goal. Packaging should be used only where needed. Proper packaging can also help prevent waste. Packaging plays an important part in preventing loss or damage to the packaged-product. Usually, the energy content and material usage of the product being packaged are much greater than that of the package. A vital function of the package is to protect the product for its intended use: if the product is damaged or degraded, its entire energy and material content may be lost.
- Minimization –The mass and volume of packaging can be measured and used as one of the criteria to minimize during the package design process. Usually "reduced" packaging also helps minimize costs. Packaging engineers continue to work toward reduced packaging.
- Reuse – The reuse of a package or component for other purposes is encouraged. Returnable packaging has long been useful for closed loop logistics systems. Inspection, cleaning, repair and recouperage are often needed. Some manufacturers re-use the packaging of the

incoming parts for a product, either as packaging for the outgoing product or as part of the product itself.

- Recycling – Recycling is the reprocessing of materials into new products. Emphasis is focused on recycling the largest primary components of a package: steel, aluminum, papers, plastics, etc. Small components can be chosen which are not difficult to separate and do not contaminate recycling operations.
- Energy recovery – Waste-to-energy and Refuse-derived fuel in approved facilities are able to make use of the heat available from the packaging components.
- Disposal – Incineration, and placement in a sanitary landfill are needed for some materials. Certain states within the US regulate packages for toxic contents, which have the potential to contaminate emissions and ash from incineration and leachate from landfill. Packages should not be littered.

Development of sustainable packaging is an area of considerable interest by standards organizations, government, consumers, packagers, and retailers.

Packaging Machines

A choice of packaging machinery includes: technical capabilities, labour requirements, worker safety, maintainability, serviceability, reliability, ability to integrate into the packaging line, capital cost, floorspace, flexibility, energy usage, quality of outgoing packages, qualifications, throughput, efficiency, productivity, ergonomics, return on investment, etc.

Packaging machines may be of the following general types:

- Blister packs, skin packs and Vacuum Packaging Machines
- Bottle caps equipment, Over-Capping, Lidding, Closing, Seaming and Sealing Machines
- Box, Case and Tray Forming, Packing, Unpacking, Closing and Sealing Machines
- Cartoning machines
- Cleaning, Sterilizing, Cooling and Drying Machines

- Converting Machines
- Conveyor belts, Accumulating and Related Machines
- Feeding, Orienting, Placing and Related Machines
- Filling Machines: handling liquid and powdered products
- Inspecting, Detecting and Check weigher Machines
- Label dispensers Help peel and apply labels more efficiently
- Package Filling and Closing Machines
- Palletizing, Depalletizing, Unit load assembly
- Product Identification: labelling, marking, etc.
- Shrink wrap Machines
- Form, Fill and Seal Machines
- Other speciality machinery: slitters, perforating, laser cutters, parts attachment, etc.

Active Packaging

The terms active packaging, intelligent packaging, and smart packaging refer to packaging systems used with foods, pharmaceuticals, and several other types of products. They help extend shelf life, monitor freshness, display information on quality, improve safety, and improve convenience.

The terms are closely related. *Active packaging* usually means having active functions beyond the inert passive containment and protection of the product. *Intelligent* and *smart* packaging usually involve the ability to sense or measure an attribute of the product, the inner atmosphere of the package, or the shipping environment. This information can be communicated to users or can trigger active packaging functions. Depending on the working definitions, some traditional types of packaging might be considered as "active" or "intellegent". More often, the terms are used with new technologically advanced systems.

Moisture Control

For many years, dessiccants have been used to actively control the moisture in a closed package. A desiccant is a hygroscopic substance usually in a porous pouch, is placed inside of a package. They have been used to reduce corrosion of machinery and to extend the shelf life of moisture sensitive foods and drugs.

Oxygen Control

Oxygen scavengers or Oxygen absorbers help remove oxygen from a closed package. Some are small packets or sachets containing powdered iron: as the iron rusts, oxygen is removed from the surrounding atmosphere. Newer systems are on cards or can be built into package films or moulded structures.

Atmosphere

With some products, such as cheese, it has long been common to flush the package with nitrogen prior to sealing: the inert nitrogen is absorbed into the cheese, allowing a tight shrink film package. The nitrogen removes oxygan and "actively" interacts with the cheese to make the package functional.

More recently, other mixtures of gas have been used inside the package to extend the shelf life. The gas mixture depends on the specific product and its degradation mechanisms. Some package components have been developed that incorporate active chemistry to help maintain certain atmospheres in packages.

Temperature Monitor

Some temperature indicators give a visual signal that a specified temperature has been exceeded. Others, Time temperature indicators, signal when a critical accumulation of temperature deviation over time has been exceeded. When the mechanism of the indicator is tuned to the mechanism of product degradation, these can provide valuable signals for consumers.

Digital temperature data loggers record the temperatures encountered throuout the shipment. This data can be used to predict product degradation and help determine if the product is suited for normal sale or if expedited sale is required. They also determine the time of the temperature excess: this can be used to direct corrective action.

Thermochromic inks are sometimes used to signal temperature excess or change. Some are reversible while others have a permanent change of color. These can be used alone or with other packaging functions such as bar codes.

The inks can also signal a desired temperature for consumers. For example, one type of beer can has ink that graphically shows when an ideal drinking temperature is achieved.

Controlling Package Temperatures

For critical vaccines, insulated shipping containers are passive packaging to help control the temperatures fluctuations seen even with a controlled cold chains. In addition, gel packs are often used to actively keep the temperature of the contents within specified acceptable temperature ranges.

Some newer packages have the ability to heat or cool the product for the consumer. These have segregated compartments where exothermic or endothermic reactions provide the desired effect.

Dispensing Systems

Some packages have closures or other dispensing systems that actively change the contents from a liquid to an aerosol. These are used for products ranging from precision inhalers for medications to spray bottles of household cleaners.

Some dispensing packages for two-part epoxy adhesives do more than passively contain the two components. When dispensed, some packages meter and mix the two components so the adhesive is fully functioning at the point of application.

RFID

Radio-frequency identification chips are becoming more common as smart labels that are used to track and trace packages and unit loads throughout distribution. Newer developments include recording the temperature history of shipments and other intelligent packaging functions.

Shock and Vibration

Shock detectors have been available for many years. These are attached to the package or to the product in the package to determine is an excessive shock has been encountered. The mechanisms of these *shock overload* devices have been spring-mass systems, magnets, drops of red dye, and several others. Recently, digital data loggers have been available to more accurately record the shocks and vibrations of shipment. These are used to monitor critical shipments to determine if extra inspection and calibration is required. They are also used to monitor the types of shocks and vibrations encountered in transit for use in package testing in a laboratory.

Other Developments

Some newer packaging films contain enzymes, anti-bacterial agents, scavengers, and other active components to help control food degradation.

Regulations

Active packaging is often designed to actively interact with the contents of the package. Thus extra care is often needed for active or smart packagings that are food contact materials. Food packagers take extra care with some types of active packaging. For example when the oxygen atmosphere in a package is reduced for extending shelf life, controls for anaerobic bacteria need to be considered. Also when a controlled atmosphere reduces the appe

Album Cover

An album cover is the front of the packaging of a commercially-released audio recording product, or album. The term can refer to either the printed cardboard covers typically used to package sets of 10" and 12" 78 rpm records, single and sets of 12" LPs, sets of 45 rpm records or the front-facing panel of a CD package, and, increasingly, the primary image accompanying a digital download of the album, or of its individual tracks.

The cover serves three main purposes:

1. To advertise and identify the contents of the music product.
2. To convey the artistic aspirations of the original artists.
3. In reproductions of the artwork, to serve as a primary image in the promotional efforts surrounding the product, as an identifiable image associated with it.

In addition, in the case of all types of records, it also serves as part of the protective sleeve.

Keep Case

A keep case or poly-box is a type of DVD packaging. From the well-known brand Amaray, this type of case is often called *Amaray case*, creating a genericized trademark and ambiguities about the real manufacturer of the case. Besides DVD-Video films, keep cases are used in most video games for the Xbox, Xbox 360, PlayStation 2, Nintendo GameCube and Wii, as well as PC titles and MP3-CD audiobooks.

Aseptic Processing

Aseptic processing is the process by which a sterile product is packaged in a sterile container in a way which maintains sterility. Sterility is achieved with a flash-heating process which retains more nutrients and uses less energy than conventional sterilization techniques such as retort or hot-fill canning. Aseptic food preservation methods allow processed food to keep for long periods of time without preservatives, as long as they are not opened.

The aseptic packages are typically a mix of paper, polyethylene, and aluminum, with a tight polyethylene inside layer. Together the materials form a tight seal against microbiological organisms, contaminants, and degradation, eliminating the need for refrigeration. The first aseptic filling plant for milk was presented in Switzerland in 1961. The Tetra Pak company later became one of the leading suppliers of processing and packaging equipment for dairies. Aseptic techniques were common in Europe and Asia before they were introduced in the United States in the 1980s.

Aseptic containers may range in size from a few fluid ounces to a nearly 8-million-gallon aseptic tank on an ocean-going ship. Aseptic processing makes worldwide export and import of new, economical and safe food products possible. Bag-In-Box technology is commonly used because it provides strong containers that are light weight and easy to handle prior to being filled. Other common package types are drink boxes and pouches. In 1991, the Institute of Food Technologists rated the top 10 innovations in food technology. Aseptic processing and packaging ranked No. 1, ahead of juice concentrates, safe canning processes, freeze-drying and food fortification. In 2007 Dr. Philip E. Nelson received the World Food Prize in recognition for his pioneering work in aseptic processing and bulk storage.

Authentication

Authentication is the act of establishing or confirming something as *authentic*, that is, that claims made by or about the subject are true.

This might involve confirming the identity of a person, tracing the origins of an artifact, ensuring that a product is what its packaging and labelling claims to be, or assuring that a computer program is a trusted one. Authentication can also be used for identity delegation. Identity Delegation in IT network is an evolving field.

Bag-In-Box

In packaging, a Bag-in-Box or BiB is a type of container for the storage and transportation of liquids. It consists of a strong bladder, usually made of several layers of metallised film or other plastics, seated inside a corrugated fiberboard box. The bag is supplied to the 'filler' as an empty pre made bag. The 'filler' then generally removes the tap, fills the bag and then replaces the tap.

The bags are available as singles for semi-automatic machines or as web bags, where the bags have perforations between each one. These are used on automated filling systems where the bag is separated on line either before the bag is automatically filled or after. There is now a technology available called FSF and pioneered by Scholle where equipment is supplied to the filler who manufactures the bags on-line from reels of film, then the FlexiTap is inserted then filled on the Scholle line.

Bag in a box packaging is liked by producers because it is inexpensive. Seen from the environmental perspective, a bag also has benefits. The bag allows a contents of 3-1000l, so that less packaging or labelling is required. The material it is made from is lighter than the alternatives, which reduces pollution caused by transport.

Battle Bag

The Battle Bag, or Mann Bag, is a lightweight load carrying system designed to allow the Combat Infantryman to carry enough ammunition and ancillaries such as medical supplies, spare batteries and radio equipment. There are a multitude of different suppliers for these, with the British Army having designed its own version for issue under the PECOC programme.

The key compartments, or pouches, on the battle bag consist of the main compartment, 3 to 4 ammo pouches, an FFD pouch, and a radio pouch, which is usually mounted within the main compartment. Some designs of battle bags have MOLLE strips on them, to allow the user to accessorise the bag to their operational requirements. Others, such as the BlackHawk bag, have an internal map case.

The Battle bag was originally designed for use as a "grab bag", to allow drivers and vehicle-borne troops to grab essential supplies in a dismount for such functions as an ambush or the assault phase of a mechanised attack. From there, it has spread in popularity with troops due to its ability to be worn comfortably over Body Armour, as well as the fact that it can be rapidly removed from the wearer in a medical emergency, or to allow the wearer to crawl through "mouseholes".

Blister Pack

Blister pack is a term for several types of pre-formed plastic packaging used for small consumer goods. The two primary components of a blister pack are the cavity or pocket made from a "formable" web, either plastic or aluminium - and the lidding, made from paper, carton, plastic or aluminium. The "formed" cavity or pocket contains the product and the "lidding" seals the product in the package.

Bottle Cap

Bottle caps, or Closures, are used to seal the openings of bottles of many types. They can be small circular pieces of metal, usually steel, with plastic backings, and for plastic bottles a plastic cap is used instead. A bottle cap is typically colorfully decorated with the logo of the brand of beverage. Caps can also be plastic, sometimes with a pour spout. Flip-Top caps like Flapper closures provide controlled dispensing of dry products. Bottle caps are often made of a different type of plastic than the bottle itself, and are often recyclable.

Bottle caps were originally designed to be pressed over and around the top of a glass bottle to grab a small flange on the bottleneck. The Crown Cork was patented by William Painter on 2 February 1892. It originally had 24 teeth and a cork seal with a paper backing to prevent contact between the contents and the metal cap. The current version has 21 teeth. To open these bottles, a bottle opener is generally advised, although some bottles incorporate a screw cap, eliminating the need for the opener.

Individuals have experimented with different ways of opening bottles, including using teeth or snapping them open across another surface. A separate interest to collecting bottle caps is that of interesting ways to open bottles. The height of the crown cap was reduced and specified in the German standard DIN 6099 in the 1960s. This also defined the "twist-off" crown cap, now widely used in the United States and Australia. This bottle cap is pressed around screw threads instead of a flange. Such a bottle cap can be taken off merely by twisting the cap.

Bottle caps are also a way for bottlers to hold promotions, especially for soda companies. A message is printed on the inside of the cap and people with the right message may win a prize. Since the bottle must be purchased to determine the message and win, people usually purchase more of the drink to increase their chances of winning. The most common prize is a free soda from that company.

Some companies, such as Snapple, also print interesting facts on the inside of their caps. Mickey's Malt Liquor as well, prints riddles underneath the 24 and 40 oz. bottle caps. Usually this is done on wide-mouthed bottles that have large caps with enough printing area to put a short sentence.

Bottle Crate

A bottle crate or beverage crate is a container used for transport of beverage containers. In the present day they are usually made of plastic, but before the widespread use of plastic they tended to be made of wood or metal. Beverage crates began to be made of HDPE in the 1950s.

Such crates can be surprisingly long-lived. In the 1980s in Sweden, a take-back campaign was organized, when 25-bottle crates were replaced by the more ergonomic 20-bottle crates. Some of the crates returned for recycling had been in use since the 1960s.

Because manufacturers avoid lead-based and cadmium-based pigmentations, in response to legislation and public opinion, they have had to resort to other methods of colouring HDPE crates. In Japan, since the early 1970s, HDPE bottle crates have been pigmented with a variety of perylene, quinacridone, azo condensation, and isoindolinone pigments. Japanese manufacturers have modified these in order to control nucleating behaviour, and have improved weathering performance and impact-resistance properties by making the light-stabilisization systems more efficient.

Box Sealing Tape

Box sealing tape or parcel tape is a pressure sensitive tape used for closing or sealing corrugated fiberboard boxes. It consists of a pressure sensitive adhesive coated onto a backing material which is usually a polypropylene or polyester film which is oriented to have strength in both the long direction and the cross direction.

Most often, the tape is 48 mm wide but it is also used in 72 mm widths. A variety of backing strengths and calipers as well as adhesive formulations are available. It is often transparent or tan. Other colors and printing are sometimes available.

Boxtop

A boxtop, in the context of being a proof of purchase, is understood to be the upper portion of a product box, detached, and mailed as part of a claim for a radio premium or other advertising offer. During the 1930s through 1950s, cereal boxtops were usually the most common proofs of purchase used to claim such premiums.

Popular cereal boxtops of the period were Wheaties, which sponsiored Jack Armstrong, the All-American Boy; Kellogg's Pep, which sponsored The Adventures of Superman; Ralston-Purina, which sponsored Tom Mix Ranston Straight Shooters; and various General Mills, particularly Cheerios and Kix, which sponsored The Lone Ranger.

Child-resistant Packaging

Child-resistant packaging or C-R packaging is special packaging used to reduce the risk of children ingesting dangerous items. This is often accomplished by the use of a special safety cap. It is required by regulation for prescription drugs, over-the-counter medications, pesticides, and household chemicals. In some juridictions, *unit packaging* such as blister packs is also regulated for child safety.

The United States Consumer Product Safety Commission has stated in a press release that "There is no such thing as child-proof packaging. So you shouldn't think of packaging as your primary line of defense. Rather, you should think of packaging, even child-resistant packaging, as your last line of defense."

Fair Packaging and Labeling Act

The *Fair Packaging and Labeling Act* is a US law that applies to labels on many consumer products.

It requires the label to state:

- The identity of the product;
- The name and place of business of the manufacturer, packer, or distributor; and
- The net quantity of contents.

The contents statement must include both metric and U.S. customary units. Passed under Lyndon B. Johnson in 1966, the law first took effect on July 1, 1967. The metric labeling requirement was added in 1992 and took effect on February 14, 1994. The law is codified as 15 U.S.C. § 1451–1461.

There has been an effort by industry threatened by a European Union directive that would force metric-only labeling starting January 1, 2010, to amend the FPLA to allow manufacturers to use metric-only labeling. Forced metric-only labeling would lead to substantial added cost due to the forced necessity of producing two distinct package types. An amendment to delay indefinitely forced metric-only labeling was adopted by the European Commission September 10, 2007, approved by the European Parliament November 29, 2007, and by the European Economic and Social Committee December 12, 2007.

Food Labeling Regulations

The law in the UK on food labeling is multifaceted and is spread over many reforms and parliamentary acts, making the subject complex. In the US, food labeling is mainly regulated by *21 CFR part 101* in accordance with *21 CFR 1.21, 74.705, Part 102, 104.20, 179.26* and FD&C. Codex Alimentarius also published a document on the food labeling which is supposed to be followed by the food industry internationally.

Nevertheless, there are general laws which should be implied on any food product:

- Name – Must also inform the customer the nature of the product. It may also be necessary to attach a description to the product name. However, there are certain generic names which must be only used for their conventional uses, for example: Muesli, Coffee, prawns.
- Ingredients – All ingredients of the food must be stated under the heading 'Ingredients' and must be stated in descending order of weight. Moreover, certain ingredients such as preservatives must be identified as such by the label 'Preservatives', a specific name, e.g. "sodium nitrite", and the corresponding European registration number colloquially known as an "E number", e.g. "E250".
- Nutritional Information– Although it is not a legal requirement to declare Nutritional information on the product, if the manufacturer makes claims that the product is 'Low in Sugar', it must be supported with nutritional information. However, as a rule it is

recommended to declare nutritional information as consumers more than ever are investigating this information before making a purchase. Moreover, there are two European nutritional labeling standards which must be adhered to if nutritional information is shown.

- Medicinal or Nutritional Claims – Medicinal and Nutritional claims are tightly regulated, some are only allowed under certain conditions while others are not authorized at all. For example, presenting claims the food product can treat, prevent or cure diseases or other 'adverse conditions' are prohibited. While claiming the food is reduced in fat or rich in vitamins require the food to meet compulsory standards and grades, in addition, the terms must be used in a form specified in regulations.
- Date Tagging – There are two types of date tagging:
  - Use by Date – 'Use by date' must be followed by a day or/and month which the product must be consumed by. To be employed on perishable foods that usually would be kept cold, for example, fish, meat, dairy products and 'ready to eat' salads.
  - Best Before Date – 'Best before date is used as an indicator of when the product will begin to degrade from optimal quality: this includes when the food becomes stale, begins to taste 'off' or decays, rots or goes mouldy. There are also regulations on which type of best before date must be applied:
    a. Best before + Day for foods with a shelf life of up to 3 months.
    b. Best before end + Month for foods with more than a 3 month shelf life.
    c. Best before end + Year for food with more than an 18 month shelf life.
- Storage Conditions – If there are any particular storage conditions for the product to maintain its shelf life, these must be pointed out. However, as a rule it is recommended to always describe the necessary storage conditions for a food product.

- Business Name and Address – In addition to the business name and address, it is necessary to indicate the manufacturer or packager, if independent to the main business and the seller established within the European Union.
- Place of Origin – The food is required to specify its place of origin, especially if the name or trademark is misleading - such as if the product is called 'English Brie Cheese' when it is produced in France.
- Instruction for Use – This is only necessary if it is not obvious how to use or prepare the product, in which case the consumer's own initiative must be used.
- Presentation – The label must be legible and easy to read, also it must be written in English, however, the manufacturer may also include other languages.
- Batch identifier, such as Lot Mark or Batch Code – It must be possible to identify individual batches with a lot mark or batch code - the code must be prefixed with the letter 'L' if it can not be distinguish from other codes, however, the date mark can be used as a lot mark. Manufacturers must bear in mind that the smaller the size of a batch, the smaller financial consequences in the case of a product recall.
- Sectioning – All of the following must be in the same field of vision:
  - Product name
  - Date mark
  - Weight
  - Quantity
  - Alcohol strength.
- Standard specification - Indicate the level of the standard compliances which the product are manufactured and packaging are completed against, and the specification limits if the standard is not publicly available, especially for those of
  - Microbial limits

  - Heavy metal limits
  - The limits of pesticide residuals
  - The limits of preservatives, artificial flavouring and colouring etc.
- Food additives - with a best practice, the items should be presented by their approved names, functional classes, and numbers of International Numbering System or equivalent.
  - Further information: United States Adopted Name
  - Further information: British Approved Name
  - Further information: E number

However, there are many other Laws and European regulations for different types of food products.

Plastic Bag

A plastic bag or pouch is a type of flexible packaging made of thin, flexible, plastics film. Plastic bags are used for containing and transporting goods such as foods, produce, powders, ice, chemicals and waste.

Most plastic bags are heat sealed together. Some are bonded with adhesives or are stitched. A press-to-close zipper can be used to open and close the bag many times.

Phamaceutical Packaging

Pharmaceutical packaging has to be carried out for the purpose of the safety of the pharmaceutical preparations in order to keep them free from contamination and hinder microbial growth in the pharmaceuticals. Packaging is one of the most important tools in the pharmaceutical industry and various pharmaceutical preparations are being used.

Packaging Gas

A packaging gas is a gas used to pack sensitive materials in a modified atmosphere. It is usually inert, or of a nature that protects the packaged goods, thus inhibiting reactions such as spoilage. Some may also serve as propellant for aerosols like whipped cream.

For packaging food, various gases are approved. Their E numbers are in parenthesis:

- Argon (E938), used for canned products
- Helium (E939), used for canned products
- Nitrogen (E941)
- Oxygen (E948), used eg. for packaging of vegetables
- Hydrogen (E949)
- Carbon dioxide (E290, also propellant)

Specific kind of packaging gases are aerosol propellants:

- Dichlorodifluoromethane known as CFC (E940), rarely used because of the damage that it does to the ozone layer
- Nitrous oxide (E942), used for aerosol whipped cream canisters
- Butane (E943a)
- Isobutane (E943b)
- Propane (E944)

Packaging Engineering

Packaging engineering, also Package engineering is a broad topic ranging from design conceptualization to product placement. All steps along the manufacturing process, and more, must be taken into account in the design of the package for any given product.

Package engineering includes industry-specific aspects of industrial engineering, marketing, materials science, industrial design and logistics. Packaging engineers must interact with Research & Development, Manufacturing, Marketing, Graphic Design, Regulatory, Purchasing, Planning and so on. The package must sell and protect the product, while maintaining an efficient, cost-effective process cycle.

Some packaging engineers have backgrounds in other science or engineering disciplines while some have college degrees specializing in this field. Formal packaging programs might be listed as package engineering, packaging science, packaging technology, etc. BS, MS, and PhD programs are available. Students in a packaging program typically begin with generalized science, business, and engineering classes before progressing into industry-specific topics such as shelf life stability, corrugated box design, cushioning, labeling regulations, distribution testing, project management, food safety, robotics, RFID tags, Quality management, package testing, packaging machinery, Tamper-evident methods, recycling, computer-aided design, etc.

Package Testing

Package testing or Packaging testing involves the measurement of a characteristic involved with packaging. This includes packaging materials, packaging components, primary packages, shipping containers, and unit loads, as well as the associated processes.

Testing measures the effects and interactions of the levels of packaging, the package contents, external forces, and end-use. It can involve controlled laboratory experiments, subjective evaluations by people, or field testing. Documentation is important: formal test method, test report, photographs, video, etc.

Testing can be a qualitative or quantitative procedure. Package testing is often a physical test. With some types of packaging such as food and pharmaceuticals, chemical tests are conducted to determine suitability of food contact materials. Testing programs range from simple tests with little replication to more thorough experimental designs.

Package testing can extend for the full life cycle. Packages can be tested for their ability to be recycled and their ability to degrade as surface litter, in a sealed landfill or under composting conditions.

Purposes

Packaging testing might have a variety of purposes, such as:

- Determine if, or verify that, the requirements of a specification, regulation, or contract are met
- Decide if a new product development program is on track: Demonstrate proof of concept

- Provide standard data for other scientific, engineering, and quality assurance functions
- Validate suitability for end-use
- Provide a basis for technical communication
- Provide a technical means of comparison of several options
- Provide evidence in legal proceedings: product liability, patents, product claims, etc
- Help solve problems with current packaging
- Help identify potential cost savings in packaging
- Predict, in a laboratory, the performance of a package during distribution and the acceptance by customers

Importance of Testing

For some types of products, package testing is mandated by regulations: food. pharmaceuticals, medical devices, dangerous goods, ect. This may cover both the design qualification, periodic retesting, and control of the packaging processes. Processes may be controlled by a variety of quality management systems such as HACCP, statistical process control, validation protocols, ISO 9000, etc. For unregulated products, the degree of package testing can be a business decision.

Risk management may involve factors such as:

- Costs of packaging
- Costs of package testing
- Value of contents being shipped
- Value of customer's good will
- Product liability exposure
- Other potential costs of inadequate packaging
- Etc

With distribution packaging, one vital packaging development consideration is to determine that a product will not be damaged throughout the entire process of getting to the customer from the manufacturer. A primary purpose of a package is to ensure the safety of a product during transportation and storage. If a product is damaged during this process, then the package has failed to accomplish its primary objective and the customer will either return the product or be unlikely to purchase the product altogether.

Package testing is often a formal part of Project management programs. Packages are usually tested when there is a new packaging design, a revision to a current design, a change in packaging material, and various other reasons. Testing a new packaging design before full scale manufacturing can save time and money.

Laboratory Affiliation

Many suppliers or vendors offer limited material and package testing as a free service to customers. It is common for packagers to partner with reputable suppliers: Many suppliers have certified quality management systems such as ISO 9000 or allow customers to conduct technical and quality audits. Data from testing is commonly shared. There is sometimes a risk that supplier testing may tend to be self-serving and not completely impartial.

Large companies often have their own packaging staff and a package testing and development laboratory. Corporate engineers know their products, manufacturing capabilities, logistics system, and their customers best. Cost reduction of existing products and cost avoidance for new products have been documented.

Another option is to use paid consultants, Independent contractors, and third-party test laboratories. They are commonly chosen for specialized expertise, for access to certain test equipment, for surge projects, or where independent testing is otherwise required. Many have certifications and accreditations: ISO 9000, ISO/IEC 17025, and various governing agencies.

Procedures

Several standards organizations publish test methods for package testing.

Included are:

- International Organization for Standardization, ISO
- ASTM International

- European Committee for Standardization. CEN
- TAPPI
- International Safe Transit Association
- etc

Governments and regulators publish some packaging test methods. There are also many corporate test standards in use. A review of technical literature and patents also provides good options to consider for test procedures. If a test is conducted with a deviation from a published test method, the test report must fully disclose that deviation.

Materials Testing

The basis of packaging design and performance is the component materials. The physical properties, and sometimes chemical properties, of the materials need to be communicated to packaging engineers to aid in the design process. Suppliers publish data sheets and other technical communications that include the typical or average relevant physical properties and the test method these are based upon. Sometimes these are adequate. Other times, additional material and component testing is required by the packager or supplier to better define certain characteristics.

When a final package design is complete, the specifications for the component materials needs to be communicated to suppliers. Packaging materials testing is often needed to identify the critical material characteristics and engineering tolerances. These are used to prepare and enforce specifications.

For example, shrink film data might include: tensile strength, elongation, Elastic modulus, surface energy, thickness, Moisture vapor transmission rate, Oxygen transmission rate, heat seal strength, heat sealing conditions, heat shrinking conditions, etc. Average and process capability are often provided. The chemical properties related for use as Food contact materials may be necessary.

Testing with People

Some types of package testing do not use scientific instruments but use people for the evaluation. The regulations for Child-resistant packaging require a test protocol that involves children. Samples of the test packages are given to a prescribed population of children. With specified 50 child panels, a high percentage must be unable to open a test package within 5 minutes. Adults are also tested for their ability to open a C-R package.

Consumer packages are often evaluated by Focus groups. People evaluate the package features in a room monitored by video cameras. The consumer responses are treated qualitatively for feedback into the new packaging process. Some food packagers use organoleptic evaluations. People use their senses to determine if a package component has tainted the food in the package.

A new package may be evaluated in a test market that uses people to try the packages at home. Consumers have the opportunity to buy a product, perhaps with a coupon or discount. Return post cards or internet sites provide feedback to package developers. Perhaps the most critical feedback is repeated sales items in the new package. Packaging evaluations are an important part of Marketing research.

Legibility of text on packaging and labels is always subjective due to the inherent variations of people. Efforts have been made to help better quantify this by people in a laboratory: still using people for the evaluation but also employing a test aparatus to help reduce variability.

Some laboratory tests are conducted but still result in an observation by people. Some test procedures call for a judgment by test engineers whether or not pre-established acceptance criteria has been met.

Package Pilferage

Pilferage is the theft of part of the contents of a package. It may also include theft of the contents but leaving the package, perhaps resealed with bogus contents. Small packages can be pilfered from a larger package such as a shipping container. Broader aspects of theft may include taking the entire package, pallet load, truck load, shoplifting, etc.

Package Insert

A package insert or prescribing information is a document

provided along with a prescription medication to provide additional information about that drug.

Oxygen Absorber

An oxygen absorber is a small packet of material used to prolong the shelf life of food. They are used in food packaging to prevent food colour change, to stop oils in foods from becoming rancid, and also retard the growth of oxygen-using aerobic microorganisms such as fungi.

Oxygen supports the growth of microorganisms and causes changes in color and rancid odors in packaged foods. Plastic packaging is less able to exclude oxygen from packaged foods than are the older glass and metal containers. Oxygen absorbers absorb oxygen and effectively reduce the aerobic environment to 0% oxygen.

Therefore aerobic bacteria and fungi are unable to grow in this environment. This will extend the shelf life of a food product from 1 week to several months. The advantages of oxygen satchels versus vacuum packaging are that the food products are not crushed or squeezed, as some products are of high value and are fragile, and its simplicity of use.

Oxygen absorbers are made in different formulations to match the water activity of the foods they are protecting. Some are designed to be used in dry foods such as nuts and dried grains, others with moister foods such as bread and processed meats.

Optical Disc Packaging

Optical disc packaging is the packaging that accompanies CDs, DVDs, and other formats of optical media. Most packaging is rigid or semi-rigid and designed to protect the media from scratches and other types of exposure damage.

Modified Atmosphere/Modified Humidity Packaging

Modified atmosphere/modified humidity packaging is a technology used to preserve the quality of fresh produce so that it can be sold to markets far away from where it is grown, extend the marketing period, and to help suppliers reduce food waste within the cold chain. Commercial examples of MA/MH include sea freight of Galia and cantaloupe melons from Central and South America to Europe and North America; transport of white asparagus from fields in Peru to markets in Western Europe; and trucking of cherries from orchards in Turkey to supermarkets in the UK.MA/MH technology is achieved by combining proprietary blends of polymers to obtain the desired MVTR and then manipulating the oxygen (O_2) and carbon dioxide (CO_2) transmission rates of the polymer by laser and/or mechanical microperforations.

The result is a film which is tailored to provide the optimum modified atmosphere and modified humidity for the produce to be packaged. "Film composition and extent of microperforation are tailored in accordance with the respiratory activity and weight of the produce packaged, anticipated temperature fluctuations during storage and shipment, and expected physiological and pathological responses of the produce to CO_2/O_2 concentrations and humidity levels inside the package.... [MA/MH] packaging allow the formation of a desirable modified atmosphere, retarding ripening and senescence of the produce.

Additional beneficial effects... include reduction of decay, chilling injury, leaf elongation, leaf sprouting, tissue discoloration, peel blemishes, and formation of off-odors, and inhibition of bacterial growth on the produce surface." Modifying the atmosphere inside fresh produce packaging to provide lowered levels of O_2 and elevated levels of CO_2 is beneficial for many fresh produce items and "can reduce respiration, decrease ethylene production and action, retard tissue ripening and softening, retard chlorophyll degradation and biosynthesis of carotenoids and anthocyanins, reduce enzymatic browning, alleviate physiological disorders and chilling injury, retard development of decay, and maintain nutritional quality of produce.

The effect of decreased O_2 and increased CO_2 on senescence and ripening process are additive and can be synergistic." "In vegetative tissues MAP can also reduce leaf regrowth, stem toughening, and leaf sprouting and rooting in root vegetables. The delay of ripening and senescence of fruits and vegetables also reduces their susceptibility to pathogens."

Maintaining high relative humidity in modified atmosphere packaging is important because it "results in reduced transpiration of water from the produce, thereby reducing wilting, shriveling, and loss of firmness. The accumulation of vapor in the packaging depends on the rate of water loss from the product, its surface area, the water vapor transmission rate of the film, and the external environment temperature."

"The RH in most sealed packaging is near the saturation level. Therefore, even very small fluctuations in temperature during storage or shipment may result in water condensation on the surface of both film and produce. Condensed water on the produce surface may adversely affect the gas exchange, leading to an unfavorable internal atmosphere." Even more detrimental may be the enhancement of produce decay, and physiological processes such as regrowth. In addition, certain human pathogens may proliferate when in-pack atmosphere contains high humidity and low O_2.

Successful delivery of fresh produce in MAP is not without potential hazards due to incompatibility of MAP film to the fruit or vegetable it is intended to keep fresh and high dependence on good temperature management. Undesirable consequences of incompatible film and/or high temperatures may be "anaerobic respiration with the accumulation of acetaldehyde, ethanol, ethyl acetate, and lactic acid, all of fermentation, contributing to the development of off odors, off flavors, and disuse deterioration."

Extreme changes in O_2 and CO_2 to intolerable levels often occur when fresh produce packed in improper MAP is exposed to higher than recommended temperatures during shipment and distribution. Other undesirable effects of improper MAP are initiation or aggravation of certain physiological disorders, irregular fruit ripening, and increased susceptibility to decay following physiological damage."

Several types of products, including retail bags, carton liners, bin bags, lidding films and flow pack have been developed from MA/MH films and are commercially available under the Xtend brand by StePac LA Ltd. Each product type is uniquely engineered for the specific produce type to be packaged. This is done through a rigorous process of developing, testing, adjusting and retesting each MA\MH product both in the lab and in commercial trials.

This meticulous process is repeated for each combination of produce and packaging type because of the many factors involved in developing a successful MA/MH packaging. These factors are "storage and shipment temperature, product respiration rate and quotient, response to levels of CO_2, O_2 and humidity, and product weight. Hence, film packaging that is adequate for consumer packages in not always suitable for bulk packaging and vice versa."

Modified Atmosphere

Modified atmosphere is an uncommon technical definition that describes the practice of modifying the composition of the internal atmosphere of a package in order to improve the shelf life. The modification process often tries to lower the amount of oxygen (O_2), moving it from 20% to 0%, In order to slow down the growth of aerobic organisms and the speed of oxidation reactions.

The removed oxygen can be replaced with nitrogen (N_2), commonly acknowledged as an inert gas, or carbon dioxide (CO_2), which can lower the pH or inhibit the growth of bacteria. Carbon monoxide can be used for keeping the red color of meat. Re-balancing of gases inside the packaging can be achieved using active techniques such as gas flushing and compensated vacuum or passively by designing "breathable" films known as equilibrium modified atmosphere packaging.

Mandatory Labeling

Mandatory labelling or labeling is the requirement of consumer products to state their ingredients or components.

Moral purchasing and problems like allergies are two things which are enabled by labelling. It is mandated in most developed nations, and increasingly in developing nations, especially for food products, *e.g.* "Grade A" meats. With regard to food and drugs, mandatory labelling has been a major battleground between consumer advocates and corporations since the late 19th century.

Because of past scandals involving deceptive labelling, countries like the United States and Canada require most processed foods to have a Nutrition facts label on the label, and the table's formatting and content must conform to strict guidelines.

The European Union equivalent is the slightly different Nutrition information table, which may also be supplemented with standardized icons indicating the presence of allergens. In China, all clothing is labelled with the factory of origin, including telephone and fax numbers, although this information is not available to buyers outside China, who see only a generic Made In China tag.

The genetic modification of food has led to one of the most persistent and divisive debates about the mandatory labelling. Advocates of such labelling claim that the consumer should make the choice whether to expose themselves to any possible health risk from consuming such foods. Detractors point to well-controlled studies that conclude genetically modified food is safe, and point out that for many commodity products, the identity of the grower and the custody chain are not known.

Voluntary labelling and co-marketing of products deemed desirable is another matter usually carried out by entirely different means, *e.g.* Slow Food. There has been increased regulatory interest in substantiating these claims, and in some jurisdictions, food labels require regulatory approval before use.

An interesting halfway is those labels that are considered mandatory by one buying population and effectively preclude purchase if they are not there, e.g. kosher, vegan, and the aforementioned GMO-free label now seen on many organic products.

Areas in which mandatory labelling is being discussed include:

- Genetically modified food
- Spraying of meat with bacteriophages

Labeling of Fertilizer

The labeling of fertilizers varies by country in terms of analysis methodology and subsequently nutrient labeling, and minimum nutrient requirements. Macronutrient fertilizers are labeled with an *NPK* analysis and also "N-P-K-S" in Australia. However, these numbers do not directly represent the source composition or absolute nutrient content of the fertilizer—macronutrient fertilizers are not labeled by content but according to traditional analysis.

For example, the fertilizer *potash* is a naturally occurring mineral composed of nearly pure potassium chloride. As such, its composition is 1:1 potassium to chloride or 52% potassium and 48% chlorine *by weight*. Traditional analysis of 100g. of *KCl* would yield 60g. K_2O. The *percentage yield* of K_2O from the original 100g. of fertilizer is the number shown on the label.

A potash fertilizer would thus be labeled 0-0-60, *not* 0-0-52. Many articles on the internet will have the wrong succession. It should always be nitrogen, phosphorus and potassium. Lawn specialists and farmers plan their entire growing season with the analysis in that order. If, for some odd reason, it was not in that order, crops and lawns could be ruined. Always look for NPK.

Fertilizer is described in this way dating back to Liebig in the 19th century due to the methods he used then to compare and evaluate fertilizers

P_2O_5 consists of 57.4% oxygen and 43.6% elemental phosphorus. The percentage of elemental phosphorus is 43.6% so P= 0.43 x P_2O_5. K_2O consists of 17% oxygen and 83% elemental potassium. The percentage of elemental potassium is 83% so K = 0.83 x K_2O.

Nitrogen values represent actual nitrogen content so these numbers do not need to be converted.

Using these conversion factors we determine that 18"51"20 fertilizer contains:

- 18% elemental (N)
- 22% elemental (P)
- 16% elemental (K)

In the U.K., fertilizer labeling regulations allow for reporting elemental mass fractions of phosphorus and potassium. The regulations stipulate this should be done in parentheses after standard *NPK* values.

Fertilizer containing only the three primary macronutrients can be referred to as *artificial* or *straight*. *Compound fertilizers* are N-P-K fertilizers with other elements purposely intermixed. For example, a mainly nitrogenous fertilizer mix would be described as a *nitrogen fertilizer*.

ISPM 15

ISPM 15 is an International Phytosanitary Measure developed by the International Plant Protection Convention that directly addresses the need to treat wood materials of a thickness greater than 6mm, used to ship products between countries. It affects all wood packaging material requiring that they be treated with heat or fumigated with methyl bromide and marked, often branded, with a seal of compliance. This seal of compliance is colloquially known as the "wheat stamp". Products exempt from the ISPM 15 are made from alternative material, like paper, plastic or wood panel products.

An ISPM 15 compliant mark must include:

- IPPC certification symbol.
- XX: represents the two letter ISO country code.
- 000: represents the unique certification number issued by AQIS to the treatment provider or wood packaging manufacturer. Inclusion of this certification number ensures that the wood packaging material can be traced back to the treatment provider and/or manufacturer.
- YY: is the treatment abbreviation where:
  - HT is the code for heat treatment to a minimum of 56o C for a minimum of 30 minutes
  - MB is the code for methyl bromide fumigation.
- ZZ is an additional code. The letters 'DB' represent debarking. DB may be added to the abbreviation of the ISPM 15 approved treatment. For example, 'HT DB' represents heat treatment and debarking and 'MB DB' represents methyl bromide fumigation and debarking.

The ISPM 15 compliant stamp may include further information as NPPOs, producers and suppliers may choose to include additional information for identification purposes.

Hermetic Seal

A hermetic seal is a seal which, for practical purposes, is considered airtight. The term is often used to describe electrical or electronic parts that are designed and intended to secure against the entry of water vapor and foreign bodies in order to maintain the proper functioning and reliability of their contents.

The food, chemical, and pharmaceutical industries all have applications for the use of barrier packaging, such as glass, Tin cans, metal foils, and high barrier plastics.

Standard test methods are available for measuring the moisture vapor transmission rate, oxygen transmission rate, etc of packaging materials. Completed packages, however, involve heat seals, joints, and closures which often reduce the effective barrier of the package. For example, the glass of a glass bottle may have an effective total barrier but the screw cap closure and the closure liner might not.

Health Claims on Food Labels

Health claims on food labels are claims by manufacturers of food products that their food will reduce the risk of developing a disease or condition. For example, it is claimed by the manufacturers of oat cereals that oat bran can reduce cholesterol, which will lower the chances of developing serious heart conditions.

Quilt Packaging

Quilt packaging is an electronic packaging technology under research that allows microchips made of dissimilar materials to be attached like squares in a quilt. Chips are connected in a way that minimizes both thermal stress and interchip delay times. The reflection coefficient s11 has been measured at less than -25dB at 40GHz in high resistivity silicon. Connections between microchips are made by thin metal lines which extend over the edge of each chip.

The current process involves the etching and deposition of silicon dioxide, evaporating and electroplating of thick copper bumps, Chemical-mechanical planarization of the copper, and the advanced silicon etch process to separate into individual chips with undercut metal nodules extending off the edges. The nodules can be joined together using ordinary lead-free soldering.

Retail Ready Packaging

Retail Ready Packaging refers to the containers and packaging for retail goods which are ready to be displayed instantly or with little set up for retail consumption by consumers.

Retail Ready Packaging is a retail industry term used by both retail stores and retail goods producers. An example of retail ready packaging is the paperboard cartons that hold several packs of gum which are placed near cash registers at supermarkets and retail stores.

Retort Pouch

A retort pouch is a type of food packaging created by aseptic processing, made from multiple layers of flexible laminate, allowing for the sterile packaging of a wide variety of food and drink, ranging from water to fully-cooked, thermo-stabilized high-caloric meals such as Meals, Ready-to-Eat which can be eaten cold, warmed by submersing in hot water, or through the use of a flameless ration heater, a meal component introduced by the military in 1992.

Retort pouches are used in field rations, space food, camping food, and brands such as Capri Sun, Tasty Bite, etc. A retort pouch is a plastic and metal foil laminate pouch that is used as an alternative to traditional industrial canning methods.

Sachet

A sachet is a small disposable bag, often used to contain single-use quantities of consumer goods, such as ketchup or shampoo. A sachet can also be a fabric bag containing pot pourri. Sale of small amounts of shampoo and detergents in plastic sachets is very popular throughout the Philippines, India and other Eastern countries.

In culinary terms, a culinary sachet is a small cloth bag made of cheesecloth, containing various herbs and spices used to infuse flavor into stocks. Mulling spice sachets containing cinnamon, cloves, allspice and orange peel are popular in the winter for flavoring ciders and wines. A plastic sachet was first used by the Cavin Kare Company to pack their shampoos.

The company failed to file a patent and hence does not own the trade rights to this invention. Scented paper envelope sachet containing scented potpourri, vermiculite and powdered fragrance compounds were introduced by Countryside Herb Farms in 1979. Sachet envelopes are used in automobiles, closets and households for aromatherapy fragrance or room deodorizer air fresheners. Sachets bags are also filled with loose spices as a simmering potpourri for stove top simmering.

Security Seal

Security seals are affixed to vehicles, chemical drums, airline duty free trolleys, utility meters and a range of boxes and various other applications, in order to provide tamper evidence and a level of security. This can help to detect theft or contamination - accidental and deliberate.

Security seals can be split into two main groups: single use and reusable. Reusable seals are usually electronic or mechanical rather than just made from plastic, and if purely plastic, only one part is reusable. *Electronic and mechanical seals* are generally used for the transportation industry, with seal fixed to the back of trucks, trailers and shipping containers. A reusable electronic or mechanical seal can be fitted to the rear or loading doors, extending a cable from itself through fixing points on the doors, and when sealed generating a unique number which is logged. When the door is next opened this number changes, so it provides tamper evidence for the load.

Single use seals are typically made of plastic or metal. The used material and production technology is the key to their security level. An indicator of a seals' security rating is whether it has obtained an ISO PAS 17712 rating. US Customs-Trade Partnership Against Terrorism and the World Customs Organisation's Framework of standards, endorsed and "encouraged" the use of ISO compliant seals.

The requirements for certain security level of seals are defined in the ISO PAS 17712 as follows:

- High security seals

- Indicative seals
- Security seals

The seals that belong to the High security level and Security level are also called *barrier seals*. They are made of different metals or metal alloys i.e. they require special tools to be open such as bolt cutters or pliers. They combine two functions - a lock and a tamper evident seal. Cable seals and bolt seals fall into this category. Designed to deter opportunistic theft of high value goods - they also can often be branded and numbered to improve tamper resistance and provide excellent security for containers. The seals that are in the Indicative level as to ISO ISO PAS 17712 are made of plastic or thin metal.

The seals can also be divided by the used material:

- Metal seals
- Plastic seals

Metal Seals

There are several basic designs of the metal seals;

- Ball seal
- Bolt seals
- Cable seals

Ball Seals

The ball seals belong to the indicative level as to ISO PAS 17712, they are used for securing trucks.

Bolt Seals

The bolt seals are used to secure shipping containers. A bolt seal used for securing containers must conform to the ISO PAS 17712 high security seal in order to be accepted by customs all around the world in ocean shipping. Especially sensitive to the conformity are the US Customs.

Cable Seals

Cable seals allow for more versatile application, as they have variable length of the locking mechanism limited only by the the cable length. Typical applications include sealing trucks and railway cars transporting valuable commodities. The cable seals are usually classified into security or high security level as to ISO PAS 17712.

Plastic Seals

There are several basic designs of the plastic seals:

- Fixed length seals
- Padlock seals
- Variable length seals

For security sealing labels, see security label.

Variable Length

Variable length seals, commonly called*pull tight seals*, can fit many applications from securing the neck of a bag or mail sack to sealing chemical drums, first aid kits and fire extinguishers. They have an adjustable length much like a cable tie but offer a much higher security level and an audit trail due to sequential numbering and the ability on some seals to be bar-coded and have a company name and logo printed or laser etched on.

Fixed Length

This type of seal tends to be more tamper resistant than variable length as there are no sliding parts, they simply 'click' into place to lock, this can make it easier to fit than a variable length seal. Most fixed length seals can also be branded and numbered for extra security. Fixed length is often the choice of single use security seals for transport applications.

Padlock Seals

Padlock seals usually are supplied as a completely plastic seal, shaped as a standard padlock would be - so can technically be classed as a fixed length seal. They feature the same types of tamper-evident protection as variable length and fixed length have available - numbering, bar-coding and branding - but fit some different applications.

The most common use for these seals is for airline duty free trolleys. On reason for this is that plastic padlocks can be placed over a locking part as any metal padlock would be to secure, but do not require keys to open, they simply break off at their built in break point using hands or sometimes a small pair of wire cutters. They are secure due to the audit trail available as a result of the sequential numbering.

Technologies for Marking the Seals

The seals can be individualized to fit the needs of the user. Typical marking includes logos, special texts, white pads for writing by hand, consecutive numbering and 1D and 2D bar codes.

There are several technologies in use:

- Hot stamping
- Ink jet printing
- Laser engraving

Self Heating Food Packaging

Self-heating food packaging is a type of active packaging that specifically refers to packaging with the ability to heat the food contents without the need for external heat sources or power. Packets typically use an exothermic chemical reaction to obtain the heat needed. Other types of active packaging are self-cooling food packaging and radio-frequency identification. Packages like these are typically used by the military during operations when it is unsafe to have a fire to cook.

Heating Methods

Commercial heat engines for self-heating food packaging involve an exothermic reaction between quicklime, or calcium oxide, and water, which generates a heat output of approximately 60 calories per gram. Quicklime, inexpensive and readily available, is generally recognized by the FDA as safe. The by-product of the reaction is calcium hydroxide.

For the military's Meals Ready to Eat or the flameless ration heater, the military currently uses a magnesium and iron heater. Water must be used to activate the heater, which generates a heat output of approximately 310 calories per gram. Hydrogen is released as a by-product. Civilian MRE versions are widely available and sold in cases. It is rarely possible to buy packets individually from legitimate sellers.

Recent Developments

Self-heating packaging continues to develop in search of inexpensive heat engines made of all food-safe materials with a reaction that emits neither odor nor fumes. One heat engine currently in development uses air-activation reactions that utilize oxidation of common metals like iron or zinc. Another uses solid fuel energy storage technology formulated to give high utilization of the chemical energy content and generates a heat output of approximately 720 calories per gram. Neither technology is commercially available.

In May, 2008, at a Product Development session of the Institute of Food Technologists convention, Michelle Richardson of the U.S. Army Natick Soldier ". In the study, dehydrated egg in a barrier laminated pouch was hydrated with integrally packaged water at time of consumption and cooked in the flameless ration heater for less than 12 minutes to produce a fresh, hot scrambled egg.

Single Use Foodservice Packaging

Single use foodservice packaging comprises disposable items often found in quick service restaurants, takeout restaurants and kiosks, and catering establishments. Typical single-use foodservice products are plates, bowls, cups, utensils, doilies, tray papers, etc. These products can be made from a number of materials including plastics, paper, bio-resins, bamboo, etc.

According to the National Restaurant Association, Americans spend nearly 50 percent of their food budget in restaurants and people therefore are using more single use foodservice packaging. Single-use packaging allows foodservice establishments to package meals in a cost-effective and sanitary fashion, while providing customers with a safe and convenient way to transport meals.

Sustainable Packaging

Sustainable packaging is the development and use of packaging which results in improved sustainability. At the end stage of design it involves increased use of life cycle inventory and life cycle assessment to help guide the use of packaging which reduces the environmental impact and ecological footprint, but in the very first steps requires a look at the whole of the supply chain: from basic function, to marketing, and then through to end of life and rebirth.

The goals are to improve the long term viability and quality of life for humans and the longevity of natural ecosystems.

Sustainable packaging must meet the functional and economic needs of the present without compromising the ability of future generations to meet their own needs.

Sustainable packaging is a relatively new addition to the environmental considerations for packaging. It requires more analysis and documentation to look at the package design, choice of materials, processing, and life cycle. This is not just the vague "green movement" that many businesses and companies have been trying to include over the past years. Companies implementing these eco-friendly actions are reducing their carbon footprint, using more recycled materials, reusing more package components, etc. They often encourage suppliers, contract packagers, and distributors to do likewise.

Environmental marketing claims on packages need to be made with caution. Ambiguous titles such as *green packaging* and *environmentally friendly* can be confusing without specific definition. Some regulators, such as the US Federal Trade Commission, are providing guidance to packagers

Companies have long been reusing and recycling packaging when economically viable. Using minimal packaging has also been a common goal to help reduce costs. Recent years have accelerated these efforts based on social movements, consumer pressure, and regulation.

Thermal Bag

A thermal bag is a type of insulated shipping container. It is usually a bag made of materials that have thermal insulation properties and is used to help maintain the temperature of any contents. Commercial thermal shopping bags carried by grocery and retail outlets, as a means to help their customers carry their temperature sensitive purchases home without breaking the cold chain, were first introduced in Europe in the mid 1980's.

Thermal bags for use with pizza were patented in 1983 by Ingrid Kosar of Gilberts, Illinois for the specific use of allowing delivered pizzas to stay hotter, longer. The first commercial application of this thermal bag was by the Domino's Pizza chain in 1984. The first retailer in the United States to offer thermal bags to their customers was Publix Supermarkets in 1998.

Tobacco Packaging Warning Messages

Tobacco packaging warning messages are health warning messages that appear on the packaging of cigarettes and other tobacco products. They have been implemented in an effort to enhance the public's awareness of the harmful effects of smoking. In general, warnings used in different countries try to emphasize the same messages. Warnings for some countries are listed below. Such warnings have been commonplace in tobacco advertising for many years. Canada was the first country in the world to adopt mandatory warning images in cigarette packages.

Vacuum Packing

Vacuum packing is a method of storing food and presenting it for sale. Appropriate types of food are stored in an airless environment, usually in an air-tight pack or bottle to prevent the growth of microorganisms. The vacuum environment removes atmospheric oxygen, protecting the food from spoiling by limiting the growth of aerobic bacteria or fungi, and preventing the evaporation of volatile components.

Vacuum packing is commonly used for long-term storage of dry foods such as cereals, nuts, cured meats, cheese, smoked fish, coffee, and crisps. It is also for storage of fresh foods such as vegetables, meats, and liquids such as soups in a shorter term because vacuum condition cannot stop bacteria from getting water which can promote their growth. Vacuum packaging food can extend its life by up to 3-5 times.

Vacuum packing is also used to reduce greatly the bulk of non-food items. For example, clothing and bedding can be stored in bags evacuated with a domestic vacuum cleaner or a dedicated vacuum sealer. This technique is sometimes used to compact household waste, for example where a charge is made for each full bag collected. Vacuum packing can be used to reduce bulk of inflatable items as well.

Vacuum packaging products using plastic bags, canisters, bottles, or mason jars are available for home use. Vacuum packaging delicate food items can be done by using an inert gas kit, typically available on chamber vacuum sealers. After air has been removed, an inert gas is added to maintain the preservation of packaged food while preventing damage. An example of inert gas for packaging delicate foods is potato chips.

Video Game Packaging

Video game packaging to the physical storage of the contents of a computer or video game, both for safekeeping and shop display. In the past, a number of materials and packaging designs were used, mostly paperboard or plastic. Nowadays, most console and computer games are shipped in jewel cases or keep cases, with little differences between them.

Aside from the actual game, many items may be included inside, such as an instruction booklet, teasers of upcoming games, subscription offers to magazines, other advertisements, or any hardware that may be needed for any extra features of the game.

Zipper Storage Bag

Zipper storage bags were patented by Robert W. Vergobbi on May 18, 1959. In the same year, Minigrip licensed them as pencil bags. Not until 1968 did Dow Chemical Company begin marketing for them. The bags are known more commonly under the genericized trademark Ziploc.

The Importance of Packaging and Labeling

Regulations for agrochemical approval vary across the globe, but packaging requirements are fairly standard.

Costs and Requirements

"Labels are very particular," says Fernando Garcia, senior product packaging engineer, crop protection, DuPont. "Every country has its own rules and regulations." Creating correct labels is vital, as improper labeling can prevent a product from being accepted. "Some things have to be in certain colors and certain sizes, and on a certain part of the package," explains Garcia.

Skimping on labels to save money can create problems. A company that doesn't need a lot of material runs into higher costs when some regulaton colors or symbols must be printed. "It becomes very expensive," says Garcia. Companies will try to get around the printing to save money, "and then, sometimes, they don't accept the labels," he says. Manufacturers do get a break on container requirements: "Country regulations differ mostly on the labeling," says Garcia. "On the packaging, they pretty much take what you offer."

However, packaging is not an area where a company wants to cut costs, as safety is paramount. "Most of the product in the chemical business for farms and agriculture is qualified as hazardous or dangerous," Garcia says. "So the importance in this industry is in an incident if you have problems with leakages. So the important thing is the materials, because of the uses."

The uses may be the same worldwide, but the distribution system in different countries can raise or lower packaging costs. For example, in the US, there are mostly big distributors. "But if you go to little markets, people buy the products in lesser quantities, so the package ends up on the shelf," Garcia says.

Some products can be packaged in bulk for export and repackaged at the retail level, saving some money, says Garcia. "But product that is more generic becomes more of a problem because every time you handle it you add cost. It's expensive to do that. So you do the finished package *in situ*, where they manufacture."

So how much of a company's budget should go toward packaging? According to Carlton Wong, director of sales and marketing and global strategy manager, Aicello North America: "Packaging costs are a variable cost; that's the percentage that you have to measure.

The normal thing is to look at the comparative price, and the metric will be what the cost of the package is per kilogram of product packaging. It becomes relative; a percentage of a total cost of sales, and that depends on how you sell the product. If you sell it in hundred grams, the package is a lot more expensive per kilo than if you sell it in five or 10 kilos."

While saving money on packaging can be difficult, investing in good product packaging will pay off in several ways. "For one thing," Garcia says, "when it's in the market and competing with other things, they want a package that displays well. Secondly, the package has to be durable and impervious to wear. It becomes an issue if you have leakage — that is a problem; that is dangerous. If you drop it and it breaks, then that becomes a big problem."

Saving Through Recycling

Some costs can be recouped through a good recycling program. "You would like to always reuse," says Garcia. Companies that package in-house can control some costs. "Reuse from the source — we use less paper, we use less plastic," says Garcia. "The best thing is returnable-reusable because you eliminate the package. You have distribution where a container comes back and then it's cleaned and refilled and goes back again."

Chemical manufacturers that don't have the luxury of producing and reusing their own packaging, or can't work with the geography and transportation required, may still get compensation through regional or national recycling programs. Brazil, for example, has an excellent recycling program where end-users simply return empty containers to a central location. "Depending on the tonnage of material that they recover, the company receives a fee back on a per-kilo basis," explains Garcia.

The Future of Packaging

As technology improves, agrochemical packagers must look at all of these factors — reducing costs, improving safety, and working toward sustainability. As technology improves, containers and packaging can become safer for both the user and the environment. Some such science already exists. "I would push more for the use of water soluble film," says Garcia. "That itself protects the package; the package doesn't become contaminated, and the product becomes the film and the product itself."

PVA film is used with water soluble packaging, "typically for powders or granules where it's intended to be a unit dose," says Wong. "It's intended to improve safety, because most accidents occur during the filling stage. So, you would package the powder, throw it into the spray tank, which would dissolve it, releasing the chemical, so there's no user contact." As EPA and other government regulations change, packaging will have to meet new requirements. DuPont is trying to stay on top of these changes, says Garcia. "We are trying to implement material that is biodegradable that disappears in 90 days, or 300 days. That's a trend that will be very important from now on. Some products need protection in the form of metallized or aluminized containers for water vapor transmission and things like that. There are efforts to get biodegradable film made of corn or soy metallized for protection of the product, and the whole thing would be biodegradable.

"Another trend that is coming up is nano-material protection," he continues, "which is made of a nanoparticles cover that would make the film with some protection for shelf life, like two or three years or so, which is very comparable to aluminum, even though it may not have the same characteristics. And that may come within the next two or three years."

As packaging costs are based on quantities, costs of these new types of containers will probably be higher when they are first introduced. However, as soon as "everyone would start using it and there is more demand, then economies of scale," says Garcia. "It certainly helps a lot more with products that command a higher price, like insecticides."

References

Arthur W. Dana.: *Kitchen Planning for Quality Food Service,* Harper Publications, New York, 1949.

Arti Sankhla, Renu Mogra and Prachi Avinash: *A Practical Guide to Food Preservation*, Agrotech Publishing Academy, Jaipur, 2014.

Ashim Gupta.: *Food and Beverage Service Management*, Centrum Press, Delhi, 2012.

2

Consumer Perceptions of Food Biotechnology

Introduction

Genetically modified foods will most likely become an increasing feature of the Asian diet in light of the region's growing demand for high volumes of quality food. The Asian Food Information Centre conducted this consumer research to provide insights on how consumers in Asia perceive the use of biotechnology to produce foods and how likely consumers are to accept the various benefits biotechnology derived foods may bring. The surveyed were conducted in 5 major cities in 5 different countries included Beijing, New Delhi, Manila, Seoul and Tokyo. This summary report focused on individual country findings - India. The other country findings will be released soon.

The purpose of this consumer research is to:

- Understand attitudes about confidence in the safety of the food supply and food labeling
- Identify concerns related to foods that have been produced using food biotechnology within the context of broader food safety and labeling issues
- Track public awareness and perceptions of food biotechnology
- Measure the extent to which direct consumer benefits of food biotechnology resonate with the public
- Uncover emerging opinions on 'sustainable food production'

Methodology

AFIC commissioned The Nielsen Company Research to

conduct a quantitative assessment of consumer attitudes towards food biotechnology. The research was conducted via an on-line survey of 1007 adults, aged 18-64, and living in 5 major cities in 5 different countries – China, India, Japan, the Philippines and South Korea. The number of respondents for New Delhi n= 204. Quotas were set to best reflect the demographic population in the cities. The survey took place from 15th July to 15th August 2008. Respondents could chose between an English language questionnaire and a version translated into the local language.

Summary of Key Findings for India

The study was divided into 3 parts: consumer attitudes towards food safety and food labeling, consumers' awareness and attitudes towards food biotechnology in general and consumer attitudes towards 'direct consumer benefits' of food biotechnology.

Confidence in the Safety of the Food Supply

Consumers in India are generally confident with the food safety. Handling/preparation and artificial ingredients/ additives are the top list of food safety concerns. In India, 19% of consumers say 'very confident' with the food safety level. More than 60% say reported to be rather/very confident. Spontaneous concerns mentioned about safety issues related to foods, in India, consumers mentioned 'Inaccurate label' and 'Hygiene' are the most concerned in food safety, followed by 'Used of low quality ingredients'.

Food biotechnology was not mentioned spontaneously as a concern. When asked to rank specific food safety concerns, 25% of Indian consumers say handling/preparation are the most concerned, followed by artificial ingredients/additives at 23%, while food biotechnology was not mentioned as a concern. At the prompted level, consumers were then shown a list of possible food safety issues and asked how concerned they were with each.

Among the five countries including India, food poisoning, pesticide residues, food additives and unknown source of food were highlighted by consumers in all countries. But Food biotechnology was not indicated as a safety concern in India.

Indian consumers seem to be alert on the food they consume with 44% claiming to avoid or eat less of certain foods: fatty/ oily food ranks highest on the list of foods consumers avoid. Biotech foods do not make the list of things consumers in India avoid.

Food Labeling

Reading food label seem to be standard practice for consumer in India. Expiry date, food additives and vitamin contents are information they are looking for. Presence of biotechnology-derived ingredients is not a labeling demand. To better understand the importance of information on food labels, respondents were questioned about their use of food labels as a source of information on nutrition and food safety.

India consumers, reading food labels seem to be standard practice, 41% say they read labels regularly and 26% say they read every single time. In all countries, expiry date is the most important information looked for on food labels, similar to India of 67%. But presence of GMO ingredients are not important information they looked for in food labeling. Indian consumers pay attention to the food they consume. When asked about information normally looked for on food labels. 97% of the India respondents say expiry date, followed by total fat content 71%, calorie/energy 58% and vitamin contents as well as food additives at 55%.

When asking about missing information on food labels, as an average, about two thirds of the India consumers, 68% are satisfied with the existing information on the labels. None of those interviewed suggested presence of biotechnology-derived ingredients as an additional item to be included on labels.

Food Biotechnology in General

India consumers have an overall positive attitude towards food biotechnology and believe that GM food will help improve food quality and food sustainability in the next 5 years. More than half of the consumers in the surveyed countries claim to have at least some knowledge on biotech food, while India, knowledge levels are slightly lower compared to the average for the region. Half of the consumers in India have limited/no knowledge. 32% of consumers think biotech foods are available in grocery shops. A better understanding of the consumer attitudes towards food biotechnology indicates the existence of two country clusters, the food production countries and the food importing countries with distinct consumer perception patterns in each. 51% Consumers in India, the food producing countries, are more positive about the benefits plant biotechnology can bring. India are favourable towards the technology and a majority of the consumers believe that food biotechnology will bring benefits in the next few years such as improved food quality, better for health and food sustainability.

Consumer Attitude Towards 'Direct Consumer Benefits' of Food Biotechnology

Majority of Indian consumers are ready to purchase biotechnology and almost all consumers support plant biotechnology if the technology is related to sustainable food production. The study was divided into nutritious foods and food sustainability. Majority of Indian consumers are ready to purchase biotechnology food such as tastier tomato, 84%, cheaper food staples, 77%, and foods/cooking oil with healthier fat profile, 71%.

When asking about food sustainability 95% support plant biotechnology if the technology is related to sustainable food production. Consumer in India ranked 'reducing the amount of pesticides needed to produce food' as the most important factors in relation to 'sustainable food production' followed by 'increasing the production of food staples in the world', thereby 'reducing world hunger'.

Conclusions

The Asian Food Information Centre provided the result of the study on how consumers in INDIA perceive the use of biotechnology to produce foods and how likely it is consumers are accepting the various benefits biotechnology derived foods may bring. India consumers seem to be the most confident with the food safety level in their countries. Their common and most important food safety concern are pesticide residues, food poisoning, and food from unknown source. Food biotechnology was not indicated as a safety concern in India. Reading food labels is the norm in India and almost 70% of the consumers are satisfied with the provided information on food labels. Expiry date is by far the most important information looked for on food labels. Food biotechnology is not a spontaneous consumer labeling demand. Half of consumers in India have at least some knowledge about biotechnology.

They have an overall positive attitude towards food biotechnology. 70% of consumers have a neutral of favourable impression of using biotechnology to produce foods and 70% strongly believe that food biotechnology will bring benefits in the next few years. Consumer in India, same as in China and the Philippines, are positive about the broad range of potential benefits biotech-derived foods can bring, expressed by a high likelihood of buying such foods. Indian consumers favour tastier food, cheaper food staples and foods/cooking oil with a healthier fat profile produced through biotechnology. A majority of consumers in the surveyed countries believe sustainable food production is important. Indian consumers rated 'reducing the amount of pesticides needed to produce food' followed by 'increasing the production of food staples in the world, thereby reducing world hunger' as the most important crop growing factors related to sustainable food production.

References

Aysha Aziz Faridi: *Dairy Microbiology*, Random Publications, Delhi, 2012.

B. K. Mishra: *Dairy and Food Processing Industry : Recent Trends (2 Vols-Set)*, Biotech Books, New Delhi, 2014.

B.D. Sharma.: *Agriculture Development and Food Security*, Ancient Publishing House, Delhi, 2012.

3

Safe Handling of Foods

Food Quality and Safety

Storage does not improve the quality of any food. The quality of a food will also not decrease significantly during storage as long as the food is stored properly and used within the recommended time frame. Quality is not the same as safety. A poor-quality food may be safe such as stale cereal, overripe fruit or soured pasteurized milk. An unsafe food may have good quality in terms of appearance and taste, but have a high bacterial count.

For example, improperly canned food may contain Clostridium botulinum thus making food unsafe. Or cooked chicken may be placed on a plate that held the raw chicken and become contaminated. The goal of home food storage is to provide both safe and high-quality foods. Maintaining a food's quality depends on several factors: the quality of the raw product, the procedures used during processing, the way the food is stored and the length of storage.

For example, fresh-picked corn will store better than corn that has been in the market for a few days; a tightly folded inside cereal box liner will prevent a ready-to-eat cereal from becoming limp. The recommended storage time takes these factors into consideration. Since bacteria frequently get into food through careless food handling, keep everything—hands, pantry, shelves and storage containers—clean.

Selection Guidelines

To help assure quality, some products have "open dates" on the package. Product dating is optional on most products. Dates may also be "coded" by the manufacturer and only understood by them.

The most commonly used open dates are:

- *Sell-by Date*: This is the last recommended day of sale. The date allows for home storage and use. You will find the date after the statement "sell by." Breads and baked goods may have "sell-by" dates.
- *Use-by Date*: Tells how long the product will retain top quality after you buy it. You will find this date after statement "Use by." Some packaged goods have "use-by" dates.
- *Expiration Date*: This is the last day the product should be used or eaten. You may find this date after the statement, "Do not use after." Yeast and baking powder have expiration dates.
- *Pack Date*: Canned or packaged foods may have pack dates, which tell you when the product was processed. This does not tell you how long the food will be good.

These are guidelines; if a food is not properly handled, its storage life will be shortened. Follow these tips for purchasing top-quality foods that have been handled safely.

- Look for packages of food that are not torn or broken.
- Canned goods should be free of dents, cracks and bulging lids.
- Refrigerated food should feel cold and frozen food should be frozen solid. Purchase these foods last.
- When shopping, place packaged raw meat, poultry and fish in plastic bags and keep from contact with other foods.
- Take perishable foods home quickly to refrigerate. If travel time will exceed one hour, pack fresh meats in a cooler with ice and keep in the passenger area of the car in warm weather.
- At home, refrigerate perishable food immediately. The "Danger Zone" for most food is 40 to 140 °F. Bacteria grow most rapidly in this range of temperatures, doubling in number in as little as 20 minutes.

Storage Guidelines

For best results in maintaining product quality practice the rule, First In, First Out. This means you use the oldest products first and the newest products later.

A good practice in the home is to place the newly purchased products in back of the same products already on the shelf. It may help to write purchase dates on products without "open dates" on the package. Follow recommended storage times for the refrigerator, freezer and pantry.

Freezer

- Keep freezer temperature at or below 0 °F. A good indication of proper temperature is that ice cream will be frozen solid.
- Use moisture-proof, freezer-weight wrap. Examples are foil, freezer bags and freezer paper. Label and date all packages.
- Food stored beyond the recommended time will be safe to eat, but eating quality and nutritive value will be reduced.
- Keep an inventory of freezer contents.

Refrigerator

- Use a thermometer to check temperature; it must be between 34 °F and 40 °F at all times. Avoid frequently opening the refrigerator door, especially in hot weather.
- Wrapping perishable food prevents the loss of flavour and the mixing of flavour and odours resulting in, for example, onion-flavoured milk.
- Raw meat and poultry should be wrapped securely so they do not leak and contaminate other foods. Place the store packages in a plastic bag or place the package on a plate to contain any juices. Clean up leaks with warm soapy water and sanitize with a solution of 1 teaspoon chlorine bleach to 1 quart water.
- Cooked meats and leftovers should be tightly wrapped to prevent leakage and drying out.

Pantry

- Storage cabinets should be cool and dry. Storage areas near oven ranges, hot water pipes or heating ducts should not be used because heat and moisture can cause a food to lose its quality more rapidly.
- High temperature or humidity may reduce storage time considerably.
- Insect infestation can occur in any home. Susceptible foods include cereals, flour, seeds, baking mixes, spices, candy, dried fruits and dry pet foods. Avoid purchasing damaged packages of foods and keep cupboard shelves clean. Storing food in tightly sealed glass, metal or rigid plastic containers may help.
- Pantry foods will probably be safe beyond recommended storage time, but eating quality and nutritive value will be reduced.

Safe Handling of Seafood

Selecting the Best

Federal regulations require that all parts of the seafood industry implement safe handling practices to keep seafood safe. The regulations require that those handling seafood have a Hazard Analysis and Critical Control Points, or HACCP plan in place in their operation. HACCP is a science-based system that examines each step in processing food, isolates possible hazards and sets limits, or critical control points, within which the food is safe.

These regulations help consumers to select safe, quality seafood. In addition, state law mandates the inspection of oyster and clam beds and other waters where seafood is harvested. The bacterial levels of these waters are inspected and are closed to the harvesting of shellfish if the levels are considered too high. Seafood poisoning can occur from eating seafood that has not been handled properly, raw or improperly cooked seafood or fish from tropical water that harbors a toxin. Consumers can protect themselves by purchasing seafood only from certified processors and dealers and by following the suggestions given here for safe handling.

Spotting a Safe Seafood Seller

Always purchase fish from a certified dealer that maintains high quality.

Based on FDA's Food Code, here are some ways of spotting a safe fish dealer:

- Employees should be in clean clothing and wearing hair coverings.
- Employees should be in clean clothing and wearing hair coverings.
- They should not be smoking, eating or playing with their hair.
- They should not be sick or have any open wounds.
- Employees should be wearing disposable gloves when handling food and change gloves after doing nonfood tasks and after handling raw fish.
- Fish should be displayed on a thick bed of fresh—not melting—ice, preferably in a case or under some type of cover. Fish should be arranged with the bellies down so that the melting ice drains away from the fish, thus reducing the chances of spoilage.

Shrimp

Fresh shrimp have a mild odour and firm-textured meat. The shell or meat is not slippery, and there are no black spots, or patches on the shell or meat. The shell of raw shrimp may be grayish-green, pinkish-tan, or light pink. When cooked, the shell turns red and the meat takes on a reddish tint. Cooked shrimp have firm meat and a mild smell.

Crabs, Lobsters and Crayfish

Live crabs, lobsters, spiny lobsters and crayfish move their legs. The "tail" of a live lobster curls under the body and does not hang down when you pick it up. Frozen spiny or rock lobster tails have clear white meat, no odour and are hard-frozen. Cooked crabs, lobsters and crayfish have bright orange to red shells and are free of any disagreeable odour.

Clams, Oysters and Mussels

Purchase raw shellfish carefully. Buy raw clams, oysters

and mussels only from reputable markets. If in doubt, ask the seafood market personnel to show you the certified shipper's tag that accompanies "shell on" products or check the shipper number on shucked oyster containers. Clams, oysters and mussels in the shell are alive, and the shells close tightly when tapped. Gaping shells indicate that the shellfish are dead and not edible. Shucked oysters are plump, and have a mild odour, a natural creamy colour, and clear liquid or nectar.

Scallops

Fresh scallops have a sweetish odour and are free of excess liquid when packaged. The meat of the large sea scallop is white, orange or pink. Smaller bay and calico scallops are white, light tan or pinkish.

Frozen Seafood

Flesh is solid, and there is no discolouration or drying on the surface. Odour is not evident or is fresh and mild. Wrapping material is moisture-and vapour-proof, fits closely around the product and is undamaged. Packaging materials do not contain ice crystals or have water stains or other indications that the product had thawed at any point.

Packaged breaded and unbreaded products have a clean and uniform appearance. Individual pieces separate easily. Breading is intact. Packaged frozen seafood may have an expiration date stamped on the label. Use the seafood before the expiration date.

Storage

Safe Handling after Purchase

Whether you've purchased seafood that is fresh or frozen, always keep it cold. Never leave perishable items in a hot car unless packed in ice or in a cooler; seafood products must be kept cold to ensure peak quality.

It's always a good idea to keep your refrigerator temperature between 32 and 38 °F, and your freezer at 0 °F or colder. Plan to use your seafood purchases within one to two days, or freeze them.

Refrigeration

Place seafood immediately in the refrigerator when you

get home from the seafood market. Wrap fresh seafood in cling wrap or store in airtight containers. Store fresh, pasteurized or smoked seafood products at 32 to 38 °F. Refrigerate live clams, oysters, mussels, crabs, lobsters and crayfish in well-ventilated containers. Cover the container with a damp cloth or paper towel.

Do not store live shellfish in airtight bags or containers. Storing live shellfish in salt water shortens their shelf life. Storing them in fresh water kills them. Keep live shellfish alive. Do not cook or eat shellfish that have died during storage.

Live clams, oysters and mussels have tightly closed shells, or the shells will close when tapped. Live crabs, lobsters and crayfish move their legs. Dead shellfish spoil rapidly and develop off-flavour and off-odours.

Freezing

Store frozen seafood products immediately in the freezer when you get home from the seafood market. Store them in their original moisture-and vapour-proof packages. Frozen seafood packaged in over wrapped trays should be repackaged in moisture-and vapour-proof plastic freezer wrap, freezer paper or foil before you store them in the freezer. Keep frozen seafood products at 0 °F or below until ready to use.

Preparation

Cleanliness

Always wash hands thoroughly with hot, soapy water before preparing foods and after handling raw seafood. Don't let raw meat or juices touch ready-to-eat foods either in the refrigerator or during preparation. Don't put cooked foods on the same plate that held raw seafood. Always wash utensils that have touched raw seafood with hot, soapy water before using them for cooked seafood. Wash counters, cutting boards and all surfaces raw seafood has touched.

Thawing

While freezing seafood quickly keeps more cell walls intact, the opposite is true for thawing. Defrost gradually so cells are disturbed less and fewer juices leak out. The best way to thaw is overnight in the refrigerator. Avoid thawing at room temperature. If you must thaw seafood quickly, here are safe options: seal seafood in a plastic bag and immerse in cold water for about an hour OR microwave on the "defrost" setting, stopping when seafood is still icy but pliable.

Marinating

Marinate seafood in the refrigerator, not on the counter. Discard the marinade after use because it contains raw juices, which may harbor bacteria. If you want to use the marinade as a dip or sauce, reserve a portion before adding raw food.

Cooking

Basic Cooking Tips for Shellfish

Shrimp, crabs, scallops, clams, mussels, oysters or lobster become tough and dry when overcooked.

To cook raw shellfish, shucked or in the shell, follow these basic guidelines:

- Raw shrimp turn pink and firm when cooked. Depending on the size, it takes from 3 to 5 minutes to boil or steam 1 pound of medium-sized shrimp in the shell.
- Shucked shellfish become plump and opaque when cooked thoroughly and the edges of the oysters start to curl. The Food and Drug Administration suggests boiling shucked oysters for 3 minutes, frying them in oil at 375 °F for 10 minutes or baking them for 10 minutes at 450 °F.
- Clams, mussels and oysters in the shell will open when cooked. The FDA suggests steaming oysters for 4 to 9 minutes or boiling them for 3 to 5 minutes after they open.
- Scallops turn milky white or opaque and firm. Depending on size, scallops take 3 to 4 minutes to cook thoroughly.
- Place lobster in a pan of boiling water to cover, and return water to boil. Boil a 1-pound lobster for 10 minutes, and allow an extra 3 minutes for each additional pound. If steaming, allow 15 to 18 minutes for a 1½-to 2-pound lobster.

Selecting and Storing Fruits and Vegetables

Selecting the Best

With satiny yellow skin and a rosy blush, it looks like the perfect peach. But how will it taste once you get it home? Choosing fresh and flavourful produce can sometimes be your greatest challenge in the supermarket. Here are some tips to find great-tasting fruits and vegetables and increase your enjoyment of these healthful foods.

Find it Fresh

With modern farming, processing and delivery, many stores are able to put produce out for sale within a day or two after it is picked. Ask your store's produce manager for delivery days so you can get to your favourite fruits and veggies before quality declines.

Select Wisely

Vegetables that are characteristic colour, shape and size generally have the best taste and texture. However, good produce doesn't have to be picture perfect. Some of the best products don't look very good. Most bananas, for example, have a fuller flavour if they are speckled.

Use your Senses

Contrary to some consumer practices, thumping or shaking a melon does not indicate ripeness. Instead, authorities recommend feeling a product. In general, produce that's too soft is too ripe; if it's too hard, it's not ripe enough. Try the sniff test, too. With certain fruits, like peaches and melons, a strong scent means they're ripening nicely.

Get the Grade

The United States Department of Agriculture (USDA) has established grade standards for most fresh fruits and vegetables. The grades are most often seen on pre-packaged apples, potatoes and onions. "U.S. Fancy" is the top grade, while "U.S. No. 1" is the most common designation. "U.S. No. 2" and "U.S. No. 3" mean lower quality.

Look for Local Produce

Fruits and vegetables grown by local farmers may be

fresher and tastier than those shipped long distances from larger farms. Once again, ask your grocery store's produce manager if any is in stock.

Go to Market

Many communities sponsor weekly farmers' markets to provide a central, in-town site for small farms to sell their produce directly to consumers.

Take a Stand

Take a weekend drive into the country to look for roadside stands where farm families sell their produce, usually picked just hours before you buy it. Or visit a farm that allows you to pick your own strawberries, blueberries, peaches and apples. Your local county Extension agent can direct you to such places.

Shop Seasonally

Probably one of the most important tips for finding great-tasting produce is to buy in season, when possible.

Here's a guide to when certain fruits and vegetables are at their peak:

- *Summer*: Apricots, blueberries, cherries, eggplant, fresh herbs, green beans, hot peppers, melon, okra, peaches, plums, sweet corn, sweet peppers, tomatoes, zucchini.
- *Fall:* Apples, broccoli, brussels sprouts, cauliflower, collards, grapes, kale, pears, persimmons, pumpkins, winter squash, yams.
- *Winter:* Beets, cabbage, carrots, citrus fruits, daikon radishes, onions, rutabagas, turnips, winter squash.
- *Spring:* Asparagus, blackberries, green onions, leeks, lettuces, new potatoes, peas, red radishes, rhubarb, spinach, strawberries, watercress.

Wax Coating

Why are wax coatings used on some fruits and vegetables? Are they safe? Many fruits and vegetables make their own natural waxy coating to help retain moisture because most produce is 80 to 95 per cent water. After harvest, but before the produce is packed and sent to the supermarket, it is repeatedly washed to clean off dirt and soil. Such extensive washing also removes the natural wax. Therefore, waxes are applied to some produce items at the packing shed to replace the natural ones that are lost.

Waxes are applied in order to:

- Help retain moisture in fruits and vegetables during shipping and marketing;
- Help inhibit mold growth;
- Protect fruits and vegetables from bruising;
- Prevent other physical damage and disease;
- Enhance appearance.

By protecting against moisture loss and contamination, wax coatings help fresh fruits and vegetables maintain wholesomeness and freshness. Waxing does not improve the quality of any inferior fruit or vegetables; rather, waxing-along with proper handling-contributes to maintaining a healthful product. Waxes by themselves do not control decay; rather, they may be combined with some chemicals to prevent the growth of mold.

The Food and Drug Administration and the Environmental Protection Agency strictly regulate the safety and use of these substances. Waxes are also used on candies, pastries and gum and come from natural sources. Wax sources generally are plants, food-grade petroleum products or insects. Some waxes can be made from dairy or animal sources, but we are not aware of any such coatings being used on fruits and vegetables in this country.

This is particularly important for people following Kosher or vegetarian diets and who don't want any animal-based wax on their produce. Any commodities that do have this type of coating must be labeled "Coated with animal-based wax." Waxes are used only in tiny amounts. In fact, each piece of waxed fruit only has a drop or two of wax. Waxes may be mixed with water or other wetting agents to ensure they are applied thinly and evenly.

The government regulates wax coatings to ensure their safety. Coatings used on fruits and vegetables must meet the food additive regulations of the U.S. Food and Drug Administration. Extensive research by governmental and scientific authorities has shown that approved waxes are safe to eat. Waxes are indigestible, which means they go through the body without breaking down or being absorbed. Produce shippers and supermarkets are required by federal law to label produce items that have been waxed so you will know whether the fruits and vegetables you buy are coated.

Consumers will see signs in produce departments that say "Coated with food-grade vegetable, petroleum, beeswax, and/or shellac-based wax or resin, to maintain freshness." None of these coatings are animal-based, and they all come from natural sources. Any consumers who have questions about wax coatings should talk to their grocers. Waxes may turn white on the surface of fruits or vegetables if they have been subjected to excessive heat and/or moisture.

This whitening is safe and is similar to that of a candy bar that has been in the freezer. Consumers do have choices. Waxes generally cannot be removed by regular washing. If consumers prefer not to consume waxes-even though the waxes are safe-they can buy unwaxed commodities or can peel the fruit or vegetable, thereby removing any coating. Commodities that may have coatings applied include apples, avocados, bell peppers, cantaloupes, cucumbers, eggplants, grapefruits, lemons, limes, melons, oranges, parsnips, passion fruit, peaches, pineapples, pumpkins, rutabagas, squash, sweet potatoes, tomatoes, turnips and yucca. However, they are not always waxed.

Table. Storage Times for Fruits and Vegetables

Food	Refrigerator	Freezer
Fruits:		
Apples	1 month	8-12 months
Apricots	3-5 days	8-12 months
Avocados	3-5 days	8-12 months
Bananas	*	8-12 months
Berries	2-3 days	8-12 months
Cherries	2-3 days	8-12 months
Grapes	3-5 days	8-12 months

Grapefruit	2 weeks	4-6 months
Guavas	1-2 days	8-12 months
Kiwis (Chinese Gooseberry)	3-5 days	4-6 months
Lemons/limes	2 weeks	4-6 months
Mangoes	*	8-12 months
Melons	1 week	8-12 months
Nectarines	3-5 days	8-12 months
Oranges	2 weeks	4-6 months
Papayas	1-2 days	8-12 months
Peaches	3-5 days	8-12 months
Pears	3-5 days	8-12 months
Pineapples	2-3 Days	4-6 months
Plantains	*	8-12 months
Plums	3-5 days	8-12 months
Rhubarb	3-5 days	8-12 months
Fruit Juices:		
Concentrate	*	2 years
Fresh or Reconstituted	5-7 days	8-12 months
Vegetables:		
Artichokes	1 week	*
Asparagus	2-3 days	8-12 months
Beets	2 weeks	8-12 months
Bok Choy	2-3 days	8-12 months
Broccoli	3-5 days	8-12 months
Brussels Sprouts	3-5 days	8-12 months
Cabbage	1-2 weeks	8-12 months
Carrots	2 weeks	8-12 months
Cauliflower	1 week	8-12 months
Celery	1 week	8-12 months
Chilies	1 week	8-12 months
Cilantro	2-3 days	*
Corn	use immediately for best flavour	8-12 months
Green Beans	1 week	8-12 months

Greens (spinach, collards, swiss chard, kale,mustard, etc.)	3-5 days	8-12 months
Jicama	2-3 weeks	8-12 months
Kohlrabi (leaves)	2-3 days	8-12 months
Kohlrabi (stems)	1 week	8-12 months
Lettuce	1 week	*
Lima Beans	3-5 days	8-12 months
Mushrooms	1-2 days	8-12 months
Onions, green	3-5 days	*
Okra	1-2 days	8-12 months
Parsley	2-3 days	*
Peas	3-5 days	8-12 months
Peppers	1 week	8-12 months
Radishes	2 weeks	*
Squash, hard	*	8-12 months
Squash, summer	3-5 days	8-12 months
Tomatillos	1 week	8-12 months
Tomatoes	1 week	8-12 months
Yuca (Cassava)	1-2 days	8-12 months
Zucchini	3-5 days	8-12 months

Notes:

* Storage by this method is not recommended due to safety or quality issues.

Selecting and Storing Cereals and Grains

Wheat Flour

Definition

Flour is the product obtained by grinding then sifting wheat kernels, or berries. The kernel consists of three distinct parts: bran, the outer covering of the grain; germ, the embryo contained inside the kernel; and endosperm, the part of the kernel containing the flour. During milling, the three parts are separated and recombined accordingly to achieve different types of flours. There are six different classes of wheat: hard red winter, hard red spring, soft red winter, hard white, soft white and durum.

The end products are determined by the wheat's characteristics, especially protein and gluten content. The harder the wheat, the higher the amount of protein in the flour. Soft, low-protein wheats are used in cakes, pastries, cookies, crackers and noodles. Hard, high protein wheats are used in breads and quick breads. Durum is used in pasta and noodles.

Storage

Flour should be stored in airtight containers in a cool, dry place. All-purpose, bread, cake and whole wheat flour will keep for six months at 90 °F, one year at 70 °F, and two years at 40 °F. Store away from foods with strong odours. Before using refrigerated or frozen flour, allow it to warm to room temperature and inspect for rancidity and taste.

Types of Flour

- White flour is the finely ground endosperm of the wheat kernel.
- All-purpose flour is a white flour milled from hard wheats or a blend of hard and soft wheats. It gives the best results for many kinds of products, including some yeast breads, quick breads, cakes, cookies, pastries and noodles. All-purpose flour is enriched and may be bleached or unbleached. Bleaching will not affect nutrient value.
- Bread flour is white flour that is a blend of hard, high-protein wheats and has greater gluten strength and protein content than all-purpose flour. Unbleached, and in some cases conditioned with ascorbic acid, bread flour is milled primarily for commercial bakers, but is available at most grocery stores.
- Cake flour is fine-textured, silky flour milled from soft wheats with a low protein content. It is used to make cakes, cookies, crackers, quick breads and some types of pastry. Cake flour has a greater percentage of starch and less protein, which keeps cakes and pastries tender and delicate.
- Self-rising flour, also referred to as phosphated flour, is a convenience product made by adding salt and leavening

to all-purpose flour. It is commonly used in biscuits and quick breads, but is not recommended for yeast breads. One cup of self-rising flour contains 1½ teaspoons of baking powder and ½ teaspoon of salt. Self-rising can be substituted for all-purpose flour by reducing salt and baking powder according to these proportions.

- Pastry flour has properties intermediate between those of all-purpose and cake flours. It is usually milled from soft wheat for pastry making, but can be used for cookies, cakes, crackers and similar products. It differs from hard wheat flour in that it has a finer texture and lighter consistency.
- Semolina is the coarsely ground endosperm of durum, a hard spring wheat with a high-gluten content and golden colour. It is hard, granular and resembles sugar. Semolina is enriched and is used to make couscous and pasta products such as spaghetti, vermicelli, macaroni and lasagna noodles. Except for some specialty products, breads are seldom made with semolina.
- Durum flour is finely ground semolina. It is enriched and used to make noodles.
- Whole wheat, stone-ground and graham flour can be used interchangeably; nutrient values differ minimally. They are produced by either grinding the whole-wheat kernel or recombining the white flour, germ and bran that have been separated during milling. Their only differences may be in coarseness and protein content. Insoluble fibre content is higher than in white flours.
- Bran is the outer layer of the wheat kernel that is removed to make white flour. It is sometimes ground into flour or it may come totally unprocessed as "miller's bran."
- Gluten flour is usually milled from spring wheat and has a high protein, low starch content. It is used primarily for diabetic breads, or mixed with other non-wheat or low-protein wheat flours to produce a stronger dough structure. Gluten flour improves baking quality and produces high-protein gluten bread.

- Wheat germ is the heart of the wheat kernel. It comes flaked or in a coarse meal, either raw or toasted. Wheat germ adds a pleasant, nutty flavour to baked goods and increases their protein and mineral content. It is the most oil-rich and therefore the most perishable part of the wheat. At the supermarket, wheat germ is found in vacuum-packed jars that must be refrigerated after opening.

Wheat Flour Terms

The Food and Drug Administration inspects and approves the use of flour treatments and additives that are used to improve the storage, appearance and baking performance of flour.

The treatments and additives are in no way harmful:

- Enriched flour is supplemented with iron and three B-vitamins and may also be supplemented with calcium. There is no change in taste, colour, texture, baking quality or caloric value of flour.
- Presifted flour is sifted at the mill, making it unnecessary to sift before measuring.
- Bromated flour is largely discontinued in the United States. Ascorbic acid is now being added to strengthen the flour for bread doughs.
- Bleached refers to flour that has been bleached chemically to whiten or improve the baking qualities. No change occurs in the nutritional value of the flour and no harmful chemical residues remain. It is only a process that speeds up the natural lightening and maturing of flour.
- Unbleached flour is aged and bleached naturally by oxygen in the air. It is more golden in colour, generally more expensive and does not have the consistency in baking qualities that bleached flour does. Unbleached is preferred for yeast breads because bleaching affects gluten strength.
- Patent flour, bleached or unbleached, is the highest grade of flour. It is lower in ash and protein with good colour. Market-wise, it is considered highest in value.

- Organic or chemical-free flour is not standardized, so its definition varies from state to state. It may be grown and stored without the use of synthetic herbicides or insecticides. It may also mean no toxic fumigants were used to kill pests in the grain, and no preservatives were added to the flour, packaging or food product.
- Gluten is a protein formed when water and wheat flours are mixed. Gluten gives bread dough elasticity, strength and gas-retaining properties. Wheat is the only grain with sufficient gluten content to make a raised or leavened loaf of bread.

Cornmeal

- Cornmeal comes in two types, either ground from yellow corn or from white. They are virtually alike, except that yellow cornmeal contains more vitamin A. Cornmeal is an "enriched" product.
- Water-ground cornmeal still contains the fat-rich germ of the corn kernel and has a fuller flavour than degerminated cornmeal. When buying, avoid packages with a rancid, stale odour.
- Blue cornmeal, from New Mexico, has a bluish-gray tint and a stronger, toastier flavour. It is often used for tortillas and tamales.
- Corn grits are also available in white or yellow. Grits are more coarsley ground than cornmeal.

Other Flour and Grain Products

- Barley flour makes a sweet, light-textured loaf of bread. It is usually combined with white flour because it does not have gluten of its own. Barley is sold also as pearl barley to be used mainly as a soup ingredient.
- Buckwheat flour is made from the seed of the fagopyrum, an herb. This flour is suitable for making pancakes. In Russia, it is called kasha. It is popular in Jewish cooking. In Japan and Korea, buckwheat noodles, called soba, are a staple.

- Oatmeal is made by rolling the groats to form flakes. Oat bran is the envelope of the groat, composed of the outer-most pericarp. It is high in soluble fibre.
- Potato flour is sometimes called potato starch and is used in baking and as a thickener. It is made from cooked potatoes that have been dried and ground.
- Quinoa has a nutty flavour, and of all grains, it comes closest to having the "perfect" protein balance.
- Rye flour is the finely ground flour obtained by sifting rye meal. The dark and light varieties are interchangeable in recipes, but the dark does have a stronger flavour. Rye and wheat flours are the only flours containing gluten-forming proteins that are important in baked goods.
- Spelt is a grain that is native to southern Europe and has a subtle, almost hazelnut flavour.
- Soy flour has a strong flavour and must be combined with wheat flour when used in baking. It is ground from whole soybeans and sometimes called soy powder or soya. Breads and cakes made with soy flour will be moist, fine-grained, and very high in iron, calcium and protein. It has none of its own starch or gluten, so it must always be combined with another flour.
- Triticale is a combination of wheat and rye and is used mostly in multigrain flours and in cereal mixes.
- Amaranth is a tall plant similar to corn, with a large shaggy head containing thousands of tiny seeds. The seeds are milled into whole grain flour, or puffed like rice or corn.

Pasta

Definition

"Pasta," an Italian word meaning paste, describes the various shapes and sizes of products made with flour and water. Pasta products may be divided into two types, dried or commercial pasta and fresh or homemade pasta.

They may also be categorized by their shapes:

- *Long Goods*: Like spaghetti

- *Short Goods*: Like macaroni
- *Specialty Products*: Like shells and bow ties
- *Noodles:* Ribbon-like shapes, generally made with eggs. By law, egg noodle products must contain 5.5 per cent egg solids by weight.

There are at least 350 shapes available in the United States and possibly 600 shapes worldwide. Pasta shapes can be used interchangeably in recipes. Pasta can be purchased plain or in a variety of colours and flavours such as tomato, spinach, herb and whole wheat, to name a few. Good-quality pasta is a golden colour, with a fine, even grain. Pasta that is too white or too gray indicates poor quality flour that will not cook properly and will be limp and sticky. The water will be cloudy after cooking.In the U.S. laws prohibit the use of artificial colouring and most pasta is enriched with iron and four B-vitamins.

Storage

Dry pasta can be stored almost indefinitely if kept in a tightly sealed package or a covered container in a cool, dry place. If cooked pasta is not to be used immediately, drain and rinse thoroughly with cold water. If the pasta is left to sit in water, it will continue to absorb water and become mushy. When the pasta is cool, drain and toss lightly with salad oil to prevent it from sticking and drying out. Cover tightly and refrigerate or freeze. Refrigerate the pasta and sauce separately or the pasta will become soggy. To reheat, put pasta in a colander and immerse in rapidly boiling water just long enough to heat through. Do not allow the pasta to continue to cook. Pasta may also be reheated in a microwave.

Preparation

Bring plenty of water to a rolling boil. Add about one tablespoon of salt per gallon of water, if desired. Add the pasta in small quantities to keep the rolling boil. Stir frequently to prevent sticking. Do not cover the pan. Follow package directions for cooking time. Do not overcook. Pasta should be "al dente". It should be slightly resistant to the bite, but cooked through. Drain pasta to stop the cooking action. Do not rinse unless the recipe specifically says to do so. For salads, drain and rinse pasta with cold water.

Rice

Most U.S. rice is enriched with iron, niacin, thiamin and folic acid. Rinsing rice, or cooking rice in excess water and draining, results in loss of enrichment and other water-soluble vitamins and minerals, and is not recommended.

Types of rice:

- *Long Grain Rice:* Long, slender kernel, four times longer than it is wide. Cooked grains are separate, light and fluffy.
- *Medium Grain Rice:* Kernel is two to three times longer than it is wide. Cooked grains are moist, tender and slightly clingy.
- *Short Grain Rice*: Kernel is almost round. Cooked grains cling together.
- *Aromatic Rice*: Brown or white rice with a natural aroma and flavour similar to that of roasted nuts or popcorn. The United States grows different kinds of aromatic rice. Some cook dry and separate and some cook moist and tender.
- *Sweet Rice*: An opaque white grain. Cooked grains are very sticky.
- *Rough Rice*: The kernels are still within the hull. Before rice can be prepackaged or cooked, the outer hull or husk must be removed.
- *Brown Rice*: Kernels of rice have had only the hull removed. Brown rice may be eaten as is or milled into regular-milled white rice. Cooked brown rice has a slightly chewy texture and a nut-like flavour. The light brown colour of brown rice is caused by the presence of bran layers.
- *Parboiled Rice*: Rough rice that has gone through a steam-pressure process before milling. Parboiled rice is favoured by consumers and chefs who desire extra fluffy and separate cooked rice.
- *Precooked Rice*: White or brown rice that has been completely cooked and dehydrated after milling. The process reduces the time required for cooking.

- *Rice Flour*: Ground from either white or brown rice. The white variety has almost no fat; the brown has a considerable amount. A variety of white rice flour called "sweet" rice flour can be used only as a thickener, so read the package label to be sure of what you are getting.

Preparation

For best results, always follow package directions. When no directions are available, use one of these methods for regular white rice.

- *Top-of-the-Range Instructions*: Combine 1 cup regular white rice, 2 cups liquid, 1 teaspoon salt and 1 teaspoon butter or margarine in 2-to 3-quart saucepan. Bring to a boil; stir once or twice. Reduce heat; cover and simmer for 15 to 20 minutes or until rice is tender. If rice is not quite tender or liquid is not absorbed, replace and cook 2 to 4 minutes longer. Fluff with fork.
- *Microwave Oven Instructions*: Combine 1 cup rice, 2 cups liquid, 1 teaspoon salt and 1 teaspoon butter or margarine in 2-to 3-quart deep microwave-proof baking dish. Cover and cook on HIGH for 5 minutes or until boiling. Reduce setting to MEDIUM and cook 15 minutes. Fluff with fork.
- *Rice Cookers*: There are several reliable brands of rice cookers available. Care should be taken to follow individual manufacturer's directions. In general, all ingredients are combined using ¼ to ½ cup less liquid than the top-of-the-range method.
- *Conventional Oven:* Cooking rice in a conventional oven with other foods is an efficient way of saving energy. Use boiling liquid. Combine ingredients in a baking dish or pan; stir. Cover tightly and bake at 350 °F for 25 to 30 minutes. Fluff with fork.
- *Reheating*: For each cup of cooked rice, add 2 tablespoons liquid. Cover and heat 4 to 5 minutes on top of range or in oven. In microwave oven, cook on HIGH about 1 to 1½ minutes per cup.

Storage

Uncooked rice can be stored on the shelf in a tightly sealed container. The shelf life of brown rice is shorter than that of white rice. The bran layers contain oil that can become rancid. Refrigerator storage is recommended for longer shelf life. Washing rice is not necessary; just measure and cook. Cooked rice can be refrigerated for up to seven days or stored in the freezer for six months.

Cereals

Corn, wheat, rice, and oats are favourite grains for making cereals. All cereals keep best in airtight containers that keep out moisture, dust and insects. At home, a tightly sealed plastic bag is sufficient protection.

Always look for a "use-by" date on the package. Crispness can be restored to ready-to-eat cereal by spreading it in a baking pan and putting it in a 350 °F oven for 5 minutes. Instant hot cereals keep up to one year in their original box, but a tightly covered container is better.

Since whole-grain cereals are rich in natural oils, purchase containers that do not have a rancid odour. Store whole-grain cereals in tightly covered containers or sealed plastic bags in the refrigerator where they should remain fresh for 5 months. When stored at room temperature, they will stay fresh one month.

Keeping Foods Safe at Home

Of the millions of cases of food borne illness that occur each year, most can be prevented.

The U. S. Department of Agriculture (USDA) advises using these strategies to keep foods safe at home:

- Buy and use a food thermometer. It's the only way to know if meat, poultry and fish are cooked safely. You can't tell just by looking.
- Use an appliance thermometer in the refrigerator and check to make sure that the temperature is 40 °F or below. In the freezer, make sure the thermometer reads 0 °F or below. Bacteria grow rapidly at temperatures above 40 °F. If the power goes out you will know what

temperatures were reached and will be able to make informed decisions about the safety of food in the refrigerator and freezer.

- Do not leave pizza sitting out on the table or a "doggie" bag in the car overnight. Foods should not be left out more than two hours at room temperature or 1 hour if it is over 90 °F. When in doubt, throw it out.
- Do not defrost a turkey in the garage or in the trunk of the car. The only safe way to defrost food is in the refrigerator, in cold water or in the microwave.
- Wash hands and all food preparation surfaces with soap and water before and after touching raw meat, poultry or fish. Bacteria on raw meat, fish or poultry can contaminate other foods such as bread or lettuce that will not be cooked.
- Do not feed "leftovers" or "take-out" food that's no longer fit for people to pets. Animals can also be stricken with food borne illnesses.
- Do not leave "take-out" or "ready-to-eat" food in the refrigerator so long that it's forgotten. You can't tell by looking at or smelling if a food is unsafe. Throw it away after three days and never taste a food when you don't know what it is or how long it has been in the refrigerator!
- Do not lick the spoon or the bowl of homemade cookie dough or cake batter made with raw eggs. Even one taste of raw dough could contain harmful Salmonella bacteria resulting in a very unpleasant and potentially dangerous illness.
- When grilling outdoors, use a clean plate for the cooked hamburgers, hot dogs or other meat or fish. Don't use the same plate that held raw meat! Juices from raw meat, poultry or fish could contaminate your cooked food.
- Separate cooked foods from uncooked foods when preparing a meal, including using separate cutting boards and knives. Cross-contamination could cause harmful bacteria from one food to be transferred to another food.

- Always put an ice pack in a child's lunch box or a lunch bag taken to the office if it includes perishable foods, such as meat, poultry, fish, milk or eggs. Foods in lunch boxes sitting in warm classrooms or offices could result in food borne illnesses. Children under the age of 10 are the most vulnerable.
- Do not "save money" by buying dented cans or cracked jars. Never use food from damaged containers. This applies to containers that are leaking, bulging or badly dented. Do not use food from cracked jars with loose or bulging lids, canned food with a foul odour or any container that spurts liquid when you open it. It's not worth taking a risk to save a few pennies.
- Put meat and poultry packages in plastic bags at the meat counter before putting them in the grocery cart. Leaking packages from meat or poultry could contaminate other foods in the cart, leading to food borne illnesses.

Food Safety Mistakes You Do Not Want to Make

Using Dirty Hands to Prepare Foods

We all know we should wash our hands before preparing food. But, did you know that not washing your hands before and during food preparation causes most food borne illnesses? Here is how to properly wash your hands: Wet your hands under warm water.

Add hand soap and scrub for 10-15 seconds before rinsing off all soap. Use a clean towel or paper towel to thoroughly dry your hands.

Cross-Contaminating

Cross-contamination is an important source of food borne illness. Cross-contamination happens when kitchen equipment is used to prepare raw meat, poultry, fish, or eggs, and then is not properly washed before preparing other foods. Harmful bacteria can transfer from these raw foods to other foods if the surface is not washed properly between uses. To prevent this, thoroughly wash any surfaces, including your hands, which come in contact with raw meat, poultry, fish or eggs.

- Clean all cutting boards, knives, utensils and counter tops with warm, soapy water before food preparation begins.
- Use one cutting board for raw meat, poultry or seafood, and a different cutting board for ready-to-eat foods, such as fresh fruits and vegetables, cheese and bread.
- Sanitize cutting boards that have been in contact with raw meat, poultry or seafood. To sanitize, put the board in a solution of 1 tablespoon liquid chlorine bleach in a gallon of warm water and leave for several minutes. Plastic cutting boards can also be sanitized in a dishwasher using the wash and dry cycle.
- To sanitize kitchen counters, first wash with hot, soapy water. Then use 1 tablespoon unscented chlorine bleach in 1 gallon of water and spread on the counter. Let sit for several minutes and dry with paper towels.
- Sanitize a non-metal kitchen sponge by heating it, while still wet, in a microwave oven for 1 to 1½ minutes. Avoid burns by allowing the sponge to cool before using it.
- Use paper towels to clean up raw meat, poultry and seafood spills on kitchen counters and other surfaces. Wash kitchen cloths and towels, which have been in contact with raw meat juices, in the hot cycle of the washing machine and dry in the dryer before reusing them.

Guessing Food is Done

Bacteria can survive on foods that are not properly cooked. Guessing if food is done by looking at changes in the colour of meat and poultry is not a good practice. Ground beef can turn brown and look done before it is safely cooked. The best way to know that food is done is to use a food thermometer. A metal-stem, digital thermometer is easy to use and removes the guesswork of when the food is done.

You can buy one at nearly any department store or grocery store:

- Clean the stem and insert into the thickest part of the food to get an accurate reading.

- Use your food thermometer to make sure the foods reach the internal temperatures shown in the chart.

These are minimum temperatures; if preferred, meats may be cooked more well-done.

Table. Minimum Internal Temperatures of Foods.

Temperature	Food
145°F	Fish steaks or fillets. All cuts of beef, lamb, pork and veal. For both safety and quality, allow meat to rest for 4 minutes before carving or eating.
155°F	Ground, mechanically tenderized or injected meats. Ground fish. Egg dishes.
165 °F	Poultry and wild game. Stuffing and casseroles.

Storing Leftovers Unsafely

Bacteria like warm temperatures especially while foods are sitting on the countertop or when they are cooling down in the refrigerator.

Refrigerate leftovers quickly so bacteria will not grow:

- As soon as you finish a meal, refrigerate your leftovers. Make sure your refrigerator is 40 °F or colder.
- Put a piece of tape on the container and write the date on the tape. It is best to use refrigerated leftovers within 4 days. If you won't be eating your leftovers in that time, put them in the freezer where they will keep safely.
- Never put a big pot of hot food in the refrigerator—it will take too long to cool to a safe temperature. Put liquid foods like hot soup or chili in shallow containers, no more than 2 inches deep. Refrigerate or freeze quickly.

Basics of Safe Food Handling

Bacteria that contaminate food and cause food borne illnesses are everywhere.

Follow these four basic safety tips to keep your food safe:

1. Wash hands and surfaces often.
2. Don't cross-contaminate.

3. Keep foods out of the temperature "Danger Zone."
4. Cook foods thoroughly.

Keep Hands and Surfaces Clean

Bacteria like Staphylococci are found on hair, skin, mouth, nose and throat. A cough or sneeze can transmit thousands of microorganisms that may cause disease. The best prevention is to keep yourself and your kitchen clean.

Keep your Hands Clean

Wash your hands! Hands become the most potentially dangerous when seemingly innocent acts like scratching the scalp, running fingers through hair, or touching a pimple become the cause for contaminating foods.

Follow the following steps to wash your hands:

- Wet hands thoroughly with warm water.
- Apply soap generously.
- Rub hands for at least 20 seconds.
- Scrub under nails with a clean nailbrush.
- Rinse hands well with warm water.
- Dry hands using a clean paper towel.

Keep Counters and Equipment Clean

Wash counters and equipment with soap and water immediately after use. Sanitize with a chlorine solution of ¾ teaspoon liquid household bleach per quart of water, especially after contact with raw meats. Use a bleach solution to sanitize the kitchen drain and disposal as well. Food particles get trapped and the moist environment is ideal for bacterial growth. Dishes and other utensils should be washed immediately in hot, soapy water and then air-dried, or cleaned in an automatic dishwasher. Bacteria can live in kitchen towels, sponges and cloths.

Wash kitchen towels and cloths regularly and always after using them to clean up raw meat juices, or use paper towels and throw them away. Wash kitchen cloths and towels in the hot cycle of the washing machine and dry them in the dryer before reusing them. Sanitize a non-metal kitchen sponge by heating it, while still wet, in a microwave oven for 1 to 1½ minutes. Avoid burns by allowing the sponge to cool before using it.

Keep Cutting Boards Clean

Whether using a wooden or plastic cutting board, it is important to keep it clean and to prevent cross-contamination after cutting raw meat, poultry and seafood. Non-porous surfaces are easier to clean than wood.

It is best to keep one cutting board for fresh produce and bread and a separate one for raw meats. This will prevent bacteria on a cutting board that is used for raw meat, poultry or seafood from contaminating a food that requires no further cooking.

Wash All Cutting Boards Thoroughly

To keep all cutting boards clean wash them with hot, soapy water after each use, then rinse and air-dry or pat dry with fresh paper towels. Non-porous acrylic, plastic or glass boards and solid wood boards can be washed in an automatic dishwasher. Laminated boards may crack and split.

Sanitize Cutting Boards Occasionally

Both wooden and plastic cutting boards can be sanitized with a solution of ¾ teaspoon liquid chlorine bleach per quart of water. Flood the surface with the bleach solution and allow it to stand for several minutes, then rinse and air dry or pat dry with fresh paper towels.

Replace Battered Cutting Boards

Even plastic boards wear out over time. Once cutting boards become excessively worn or develop hard-to-clean grooves, they should be discarded.

Prevent Cross-Contamination

Cross-contamination is the transportation of harmful substances to food by:

- Hands that touch raw foods, such as raw meat, then touch food that will not be cooked, like salad ingredients.
- Surfaces or cleaning cloths that touch raw foods, are not cleaned and sanitized, then touch ready-to-eat food.

- Raw meat, raw poultry and raw seafood that touch or drip fluids on cooked or ready-to-eat foods.

Keep Foods out of the "Danger Zone"
Safely Store Perishable Foods

Refrigerate or freeze foods that will spoil at room temperature. Keep your refrigerator between 34 °F and 40 °F and your freezer temperature at or below 0 °F. The "Danger Zone" for most foods is between 40 °F and 140 °F. Bacteria grow most rapidly in this range of temperatures, doubling in number in as little as 20 minutes. Discard any perishable food left out at room temperature for more than two hours.

Safely Thaw Foods

Thaw and marinate foods in the refrigerator, never on the counter. If thawed at room temperature, bacteria can grow in the outer layers of the food before the inside thaws. Proper thawing is essential to maintaining the safety, taste and texture of frozen foods.

It affects the juiciness of meats, the texture and flavour of vegetables and fruits, and moisture level of baked goods:

- Thick meat cuts should be thawed before cooking to retain juiciness. Cuts such as chops, patties and steaks that will be pre-pared by flouring or breading should be thawed before baking.
- Broccoli, cauliflower and greens are more flavourful if partially thawed before cooking.
- Thawing foods should be placed in a shallow pan to catch drippings so that other refrigerated foods will not be contaminated with raw food juices.\
- Never thaw foods at or above room temperature. Remember food spoilage bacteria multiply most rapidly at temperatures between 40 °F and 140 °F.
- Thaw frozen fruits, vegetables or meat in the refrigerator overnight, in a sealed freezer container. Foods may be thawed more quickly by immersing the sealed freezer container into cold water and changing the water frequently until food is thawed. Foods may also be thawed in the microwave using the defrost setting.

- If thawed in the microwave or in cold water in the sink, food must be cooked immediately after thawing. Do Not thaw a food and then refrigerate to cook later.
- When you have defrosted food for use, keep in mind that thawed frozen food is more perishable than fresh food.
- Thawed foods that have been at room temperature for over two hours should be discarded.
- Foods thawed in the refrigerator may be refrozen IF they still contain ice crystals. Immediately remove only the amount needed from the freezer container, remove air, reseal and return remaining food to the freezer.
- Thawed meats and poultry kept in the refrigerator should be used within two to three days. Thawed seafood kept in the refrigerator should be used within one to two days.
- Thaw bread and baked goods at room temperature in sealed freezer containers or original wrapping to avoid moisture loss.

Table. Recommended Times for Refrigerator and Freezer Food Storage.

Food	Refrigerator	Freezer
Dairy		
Fresh milk	5-7 days	*
Buttermilk	1-2 weeks	*
Canned Milk (opened)	3-5 days	*
Yogurt, cottage cheese	7 days	*
Hard cheese	6-12 weeks	6-12 months
Cheese spreads	3-4 weeks	*
Ice cream	*	2 months
Eggs		
Fresh in shell	3 weeks	*
Hard-cooked	1 week	*
Meats, Fresh		
Beef roasts, steaks	3-5 days	6-12 month

Ground beef or stew	1-2 days	3-4 months
Pork roast, chops	3-5 days	4-6 months
Sausage	1-2 days	1-2 months
Chicken or turkey	1-2 days	9-12 months
Meats, Cooked		
Smoked Sausage, whole ham (fully cooked)	7 days	1-2 months
Ham slices (fully cooked)	3-4 days	1-2 months
Hotdogs, luncheon meats (unopened)		
Hotdogs, luncheon meats (opened)	3-7 days	1-2 months
Leftover meat, cooked	3-4 days	2-3 months
Leftover gravy and meat broth	1-2 days	2-3 months
Leftover poultry, cooked	3-4 days	4-6 months
Seafood		
Fresh lean fish: cod, flounder, trout,	1-2 days	4-6 months
haddock, halibut, pollack, perch		
Fresh fatty fish: mullet, smelt, salmon,	1-2 days	2-3 months
mackerel, bluefish, tuna, swordfish		
Live crabs and lobster	same day purchased	* *
Live mussels and clams	2-3 days	*
Live oysters	7-10 days	*
Freshly shucked oysters	5-7 days	3-4 months
Scallops, shrimp, shucked mussels and clams	2-3 days	3-4 months
Fruits and Vegetables (Fresh)		
Apples	1 month	8-12 months
Apricots, avocados, grapes, peaches,	3-5 days	8-12 months
pears, plums		

Berries, cherries	2-3 days	8-12 months
Grapefruit, lemons, limes, oranges	2 weeks	4-6 months
Pineapple	2-3 days	4-6 months
Beets, carrots	2 weeks	8-12 months
Beans, broccoli, greens, peas, summer squash	3-5 days	8-12 months
Celery, cabbage, chilies, lettuce, peppers, tomatoes	1 week	8-12 months
Mushrooms	1-2 days	8-12 months
Pies		
Chiffon pie, pumpkin pie	1-2 days	1 month
Fruit pie	1-2 days	1 year

Notes:

* Storage by this method is not recommended due to safety or quality issues.

Cook Foods Thoroughly

Using a thermometer is the only reliable way to ensure safety and to determine the "doneness" of meat and egg dishes. To be safe, these foods must be cooked to an internal temperature high enough to destroy any harmful bacteria that may have been in the food. Colour changes in meat are no longer considered reliable proof that all bacteria have been destroyed. Use the following minimum internal temperature chart to determine if foods have been cooked thoroughly.

Minimum Internal Temperatures

These temperatures ensure that food borne bacteria have been destroyed. For reasons of personal taste or texture preferences, consumers may choose to cook meat and poultry to higher temperatures.

Table. Minimum Internal Temperatures of Foods.

Temperature	Food
145 °F	Fish steaks or fillets. All cuts of beef, lamb,pork and veal. For both safety and quality, allow meat to rest for 4 minutes before carving or eating.

155°F	Ground, mechanically tenderized or injected meats. Ground fish. Egg dishes.
165°F	Poultry and wild game. Stuffing and casseroles.

Safely Handle Leftovers

Divide large amounts of hot leftovers directly into small, shallow containers for quick cooling, and place directly in the refrigerator. Discard food that has been left standing at room temperature for more than two hours. Date leftovers so they can be used within a safe time.

Most foods remain safe when refrigerated for three to five days. If you will not be eating the leftovers within that time, freeze them for longer storage. If in doubt, throw it out rather than risk a food borne illness. Never taste food that looks or smells strange to see if you can still use it. Even a small amount of contaminated food can cause illness.

Effects of Temperature on Food

Background

"Last night I left cooked roast beef on the counter to cool before refrigerating but fell asleep and discovered it this morning. I immediately put it in the refrigerator. Since the meat is cooked, shouldn't it be safe to eat?" The answer to this question is that the roast beef should be thrown out.

Why? Because leaving food out too long at room temperature can cause bacteria-such as Staphylococcus aureus, Salmonella enteriditis, Escherichia coli O157: H7, and Campylobacter-to grow to dangerous levels that can cause illness. Bacteria exist everywhere in nature.

They are in the soil, air, water and the foods we eat. When the bacteria have nutrients, moisture, time and favourable temperatures, they grow rapidly increasing in numbers to the point where some can cause illness. Therefore, understanding the important role temperature plays in keeping food safe is critical. If we know the temperature at which food has been handled, we can then answer the question, "Is it safe?"

The Danger Zone

Bacteria grow most rapidly in the range of temperatures

between 40 and 140 °F, doubling in number in as little as 20 minutes. This range of temperatures is often called the "Danger Zone." That's why perishable foods should never be left out of refrigeration over two hours. If the temperature is above 90 °F, food should not be left out more than one hour.

Cooking

Raw meat and poultry should always be cooked to a safe internal temperature. When roasting meat and poultry, an oven temperature no lower than 325 °F should be used. Use a meat thermometer to assure that meat and poultry have reached a safe internal temperature. Cook beef, lamb, pork and veal steaks, roasts and chops to a safe minimum internal temperature of at least 145 °F, then allow to rest for 4 minutes before carving or eating.

Cook longer if you prefer your meat more well-done. Cook ground meats, mechanically tenderized, or injected meats to an internal temperature of 155 °F. All poultry is safe if it reaches a minimum internal temperature of 165 °F throughout the product. For reasons of personal taste preferences, poultry may be cooked longer. If raw meat and poultry have been handled safely, using the above preparation recommendations will make them safe to eat. If raw meats have been mishandled, bacteria may grow and produce toxins, which can cause food borne illness. Cooking does not destroy toxins that are heat-resistant. Therefore, even though cooked, meat and poultry mishandled in the raw state may not be safe to eat even after proper preparation.

Storing Leftovers

One of the most common causes of food borne illness is improper cooling of cooked foods. Bacteria are everywhere, even after food is cooked to a safe internal temperature, and they can be reintroduced to the food and then reproduce. For this reason leftovers must be put in shallow containers for quick cooling, and refrigerated immediately or within two hours of preparation.

Reheating

Foods should be reheated thoroughly until hot and steaming, and to an internal temperature of 165 °F. In the microwave oven, cover food and rotate so it heats evenly. Follow manufacturer's instructions for stand time for more thorough heating. In the absence of manufacturer's instructions, at least a two-minute stand time should be allowed.

Cold Storage Temperatures

Properly handled food stored in a freezer at 0 °F will always be safe. Freezing keeps food safe by slowing the movement of molecules, causing bacteria to enter a dormant stage. Once thawed, these bacteria can again become active and multiply to levels that may lead to food borne illness. Because bacteria on these foods will grow at about the same rate as they would on fresh food, thawed foods should be handled as any other perishable food.

A temperature of 34 to 40 °F should be maintained in the refrigerator. In contrast to freezer storage, perishable foods will gradually spoil in the refrigerator. Spoilage bacteria will make themselves known in a variety of ways. The food may develop an uncharacteristic odour, colour and/or become sticky or slimy.

Molds may also grow and become visible. Bacteria capable of causing food borne illness either don't grow or grow very slowly at refrigerator temperatures. A refrigerator/freezer thermometer should always be used to verify that the temperature of the unit is correct. Safe food-handling practices are a good defence against food borne illness. Because we know how different temperatures affect the growth of bacteria in our food, we can protect ourselves and our families from food borne illnesses by proper handling and cooking, and by storing foods at safe temperatures.

Safe Handling of Beef

Selecting the Best

Fresh is Best

Choose beef that is bright red in colour from the fresh meat case. Vacuum packaged beef will be maroon because it has not been exposed to oxygen. Choose beef that is firm to the touch. Look for packages that are cool to the touch, have no wear or punctures, and little or no excess liquid. Always check the "sell-by" date. "Loin" and "rib" are clues that the beef is a more tender cut. "Chuck", "round" and "flank" indicate a less tender cut. Less tender cuts will require marinating or a slower, moist cooking method. Purchase raw meats last.

Make sure all meats, whether raw, pre-packaged or from the deli, are refrigerated when purchased. Fresh meats may contaminate other grocery items. The best way to prevent this cross-contamination is to always keep fresh meats from other items. Put raw meat packages in a plastic bag so juices won't drip onto other foods. Pack raw meats in an ice chest if it will take you more than an hour to get home, and keep the ice chest in the passenger area of the car during warm weather. Take meats straight home to the refrigerator or freezer.

Product Dating

Product dating, applying "sell by" or "use by" dates, is not required by federal regulations. However, many stores and processors may voluntarily choose to date packages of raw beef. Use or freeze products with a "sell by" date within three to five days of purchase. If the manufacturer has determined a "use by" date, observe it. It's always best to buy a product before its date expires. It's not important if a date expires after freezing beef, because all foods stay safe while properly frozen.

Product Inspection

All beef found in retail stores is either USDA inspected for wholesomeness or inspected by state systems which have standards equal to the federal government. Each animal and its internal organs are inspected for signs of disease. The "Passed and Inspected by USDA" seal ensures that the beef is wholesome and free from disease. Although inspection is mandatory, its grading for quality is voluntary, and a plant pays to have its beef graded.

Product Grading

Quality Grade refers to the eating quality of the meat. Grades are determined by the amount of marbling, the texture of the meat and its colour and appearance.

USDA grades for beef are as follows:

- *USDA Prime:* The lean is highly marbled and usually very tender and juicy; outside fat may be excessive.

- *USDA Choice*: The lean is average in marbling and usually tender and juicy; outside fat is variable.
- *USDA Select*: The lean contains some marbling; tenderness and juiciness can be extremely variable; usually not much outside fat.
- *USDA Standard:* Little or no marbling; tenderness and juiciness extremely variable; very little outside fat.
- *USDA Commercial, Utility, Cutter and Canner*: Generally applied to older animals. This beef is most often used in processed products and is rarely cut for the freezer.

Storage

Refrigeration

Keep beef below 40 °F during storage. Store uncooked beef items together, separate from cooked foods. Refrigerate or freeze fresh beef Immediately after bringing it home. Never leave beef in a hot car or sitting out at room temperature. Packaged whole cuts of fresh beef may be refrigerated in their original wrappings in the coldest part of the refrigerator for three to five days after purchase, while ground beef can be stored in the refrigerator for one to two days. Keep beef refrigerated until you are ready to cook it. When transporting cooked beef to another dining site, place it in an insulated container or ice chest until ready to eat. Cooked whole cuts of beef are best when refrigerated no longer than two to three days. Cooked ground beef is best when refrigerated no longer than one to two days.

Freezing

Freeze whole cuts of fresh beef if you do not plan to cook it within three to five days after purchase. Freeze ground beef if you do not plan to cook it within one to two days after purchase. It is safe to freeze beef in its original packaging or repackage it. However, for long-term freezing, overwrap the porous store plastic with aluminum foil, freezer paper, or place in freezer bags to prevent "freezer burn", which appears as grayish brown leathery spots and is caused by air reaching the surface of food. For best quality, use steaks and roasts within 6 to 12 months, ground beef within 3 to 4 months.

Preparation

Cleanliness

Always wash hands thoroughly with hot soapy water before preparing foods and after handling raw beef. Don't let raw meat or juices touch ready-to-go foods, either in the refrigerator or during preparation. Don't put cooked foods on the same plate that held raw beef. Always wash utensils that have touched raw meat with hot, soapy water before using them for cooked meats. Wash counters, cutting boards and other surfaces raw meats have touched. These surfaces may be sanitized by cleaning with a solution of 1 teaspoon chlorine bleach per quart of water.

Thawing

Thaw uncooked beef in the refrigerator or in cold water. Never thaw beef at room temperature. Thawing by refrigeration requires planning ahead and most likely allowing a 24-hour thawing period. To thaw beef in cold water, leave the beef in its original wrapping or place it in a watertight plastic bag.

Change the water every 30 minutes. For quick thawing of uncooked or cooked beef, use the microwave, but plan on cooking the meat immediately after thawing because some areas of the food may become warm and begin to cook during microwaving. Thawing time will vary according to whether you're thawing a whole roast or cuts and the number of parts frozen together. Use the Defrost or Medium-Low setting, according to the manufacturer's directions. Turn the roast and separate parts as they thaw, taking care the meat does not begin to cook. Repeat as needed. Foods defrosted by the cold water method or in the microwave should be cooked before refreezing, because they may have been held at temperatures above 40 °F.

Table. Storage Times for Beef Products.

Products	Refrigerator (40 °F)	Freezer (0 °F)
Fresh beef roast, steaks, chops, or ribs	3 to 5 days	6 to 12 months
Fresh beef liver or variety meets	1 to 2 days	3 to 4 months

Home cooked beef, soups, stews or casseroles	3 to 4 days	2 to 3 months
Store-cooked convenience meals	1 to 2 days	2 to 3 months
Cooked beef gravy or beef broth	1 to 2 days	2 to 3 months
Beef hot dogs or lunch meats, sealed in package	2 weeks (or 1 week after a "Use by" date)	1 to 2 months
Beef hot dogs, opened package	7 days	1 to 2 months
Lunch meats, opened package	3 to 5 days	1 to 2 months
TV dinners, frozen casseroles	Keep only in freezer	3 to 4 months
Canned beef products in pantry	2 to 5 years in pantry; 3 to 4 days after opening	After opening, 2 to 3 months
Jerky, commercially vacuum packaged	1 year in pantry; 2 to 3 months in refrigerator	Do not freeze

It is safe to cook frozen beef in the oven, on the stove, or grill without defrosting it first, although the cooking time may be about 50% longer. Do not cook frozen beef in a slow cooker.

Marinating

Marinate food in the refrigerator, not on the counter. If you want to use the marinade as a dip or sauce, reserve a portion before adding raw food. Discard uncooked leftover marinade, or bring to a full boil before brushing on cooked beef.

Partial Cooking or Browning

Never brown or partially cook beef, then refrigerate and finish cooking later, because any bacteria present would not have been destroyed. It is safe to partially pre-cook or microwave beef immediately before transferring it to a hot grill or oven to finish cooking.

Cooking

Importance of Kitchen Thermometers

One of the critical factors in controlling bacteria in food is controlling temperature. Pathogenic microorganisms grow very slowly at low temperatures, multiply rapidly in mid-range temperatures, and are killed at high temperatures. Cook foods thoroughly to prevent food borne illness. Using a thermometer is the only reliable way to ensure safety and to determine the "doneness" of beef and most other foods. To be safe, a product must be cooked to an internal temperature high enough to destroy any harmful bacteria that may have been in the food. Many food handlers believe that visible indicators, such as colour changes in the food, can be relied on to determine whether foods have been cooked long enough to ensure bacterial destruction. However, recent research has shown that colour and texture indicators are not reliable.

Whole Muscle Meats

When cooking whole cuts or parts of beef, the thermometer should be inserted into the thickest part of the meat, away from the bone, fat and gristle. The thermometer may be inserted sideways if necessary.

When the food being cooked is irregularly shaped, the temperature should be checked in several places. Steaks and roasts should be cooked to 145 °F followed by a 4 minute rest, 160 °F, or 170 °F.

Ground Beef

Ground beef must be cooked thoroughly to kill harmful bacteria. Unlike whole muscle meat, whose interior meat is sterile, the grinding process exposes the interior meat in ground beef to bacteria, which may be on the surface, in the air, on equipment or on people's hands. To kill these bacteria, food safety experts have one major rule of thumb—cook ground beef to at least 155 °F. This step, while very simple, offers the best protection that consumers can have for serving ground beef products safely.

Microwaving

When microwaving unequally sized pieces of beef, arrange

them in a dish or on a rack so thick parts are towards the outside of the dish and thin parts are in the center, and cook on medium-high or medium power. Place a roast in an oven-cooking bag or in a covered pot. Refer to the manufacturer's directions that accompany the microwave oven for suggested cooking times. Remove from microwave and test for doneness in several places with a meat thermometer.

Serving

Basic Tips

Wash hands with soap and water before serving or eating food. Serve cooked products on clean plates with clean utensils and clean hands. Never put cooked foods on a dish that has held raw products unless the dish is washed with soap and hot water. Hold hot foods above 140 °F and cold foods below 40 °F. Never leave foods, raw or cooked, at room temperature longer than two hours. On a hot day with temperatures at 90 °F or warmer, this decreases to one hour.

Leftovers

Basic Tips

Always use clean utensils and storage containers for safe storage. Divide large amounts of leftovers into small, shallow containers for quick cooling in the refrigerator; avoid placing large pots of gravy or stew in the refrigerator to cool since it will likely take until the next day for this amount of food to cool.

To store in the refrigerator, wrap cooked meat in plastic wrap or aluminum foil, or store it in a tightly covered container and use within two to three days. For frozen storage, wrap meat in freezer paper, heavy aluminum foil, or place in freezer bag, and use within two to three months. Never taste food that looks or smells strange to see if you can still use it. If In Doubt Throw It Out!

Table. Approximate* Beef Cooking Times.

Type of Beef	Size	Cooking Method	Cooking Time	Internal Temperature
Rib roast, bone-in	4 to 6 lbs.	Roast 325 °F	23 to 25 min./lb.	Medium rare 145 °F

Rib roast, boneless rolled	4 to 6 lbs.	Roast 325 °F	Add 5 to 8 min./lb. to times above	Medium rare 145 °F
Chuck roast, brisket	3 to 4 lbs.	**Braise 325 °F	2 to 3 hours	Medium 160°F
Round or rump roast	2½ to 4 lbs.	Roast 325 °F	30 to 35 min./lb.	Medium rare 145 °F
Tenderloin, whole	4 to 6 lbs.	Roast 425 °F min. total	45 to 60 145 °F	Medium rare
Steaks	¾" thick	Broil/Grill per side	4 to 5 min. 145 °F	Medium rare
Stew or shank cross cuts	1 to 1½" thick	Cover with liquid; simmer	2 to 3 hours	Medium 160 °F
Short ribs	4" long and 2" thick	**Braise 325 °F	1½ to 2½ hours	Medium 160 °F

Notes:

*For safety, cook hamburgers and ground beef mixtures such as meat loaf to 155 °F on a meat thermometer. However, whole muscle meats such as steaks and roasts may be cooked to 145 °F (medium rare), if that is followed by a 4 minute rest before carving or eating. The cooking times listed here are approximate for use in meal planning. Use a meat thermometer to check for safe cooking and doneness of beef.

**Braising is roasting or simmering less-tender meats with a small amount liquid in a tightly covered pan.

Safe Handling of Canned Goods

Canning is an important, safe method of food preservation if practiced properly.The canning process involves placing foods in jars or cans and heating them to a temperature that destroys micro-organisms that could be a health hazard or cause the food to spoil. Canning also inactivates enzymes that could cause the food to spoil.

Air is driven from the jar or can during heating, and as it cools, a vacuum seal is formed. This vacuum seal prevents air from getting back into the product bringing with it microorganisms to recontaminate the food.

Purchasing Canned Goods

Buy cans and jars that look perfect. Carefully check dented cans and jars for leakage and rust before buying. Cans and jars should be free of dents, cracks or bulging lids, which can indicate a serious food poisoning threat. Dusty cans or torn labels may indicate old stock. Also check for pull dates.

Product Dating

Dates on product packages recommend purchase or "use by" dates. They are not safety dates. Product dating is voluntary and not required by federal regulations. Since product dating is voluntary by federal regulations, a retailer may legally sell foods beyond the date on the package as long as the product is wholesome. However, it is not legal to alter, change or cover up a date on a product packaged under federal inspection. The product may continue to be offered for sale but the expired date must remain visible to the consumer.

Types of Dates:

- A "sell by" date tells the store how long to display the product for sale. You should buy the product before the date expires to have sufficient time to use it at best quality.
- A "best if used by" date is recommended for best flavour or quality. It is not a purchase or safety date.
- A "use by" date is the last date recommended for the use of the product while at peak quality and has been determined by the manufacturer of the product.
- "Closed or coded dates" are packing numbers for use by the manufacturer to rotate the stock as well as to locate their products in the event of a recall.

Storage of Canned Goods

Storage does not improve the quality of any food. The quality of a food will not decrease significantly if stored properly and if the food is eaten within the recommended time frame. For best results in maintaining product quality, practice the rule, First In, First Out. This means you use the oldest products first. A good practice in the home is to place the newly purchased cans in back of the same products already on the shelf.

For best quality, use home canned foods within one year, and commercially processed cans within two years. Storage cabinets should be cool and dry. The best temperature for storing canned foods is between 50 °F and 70 °F. Avoid storing canned foods in a warm place near hot pipes, a range or furnace, or in direct sunlight. Storage time decreases significantly when temperatures are above 75 °F. Keep canned goods dry to prevent cans or metal lids from rusting, which may cause cans to leak and food to spoil.

Recommended Storage for Unopened Canned Foods in a Cool, Dry Pantry:

- *High-Acid Canned Food:* 12 to 18 months Juices, tomatoes, grapefruit, apple products, mixed fruit, berries, pickles, sauerkraut and vinegar-based products.
- *Low-acid Canned Foods:* 2 to 5 years Meat and poultry products, vegetable soups, spaghetti products, potatoes, corn, carrots, beans, beets, peas, pumpkins, etc.
- *Home-Canned Foods:* All types 1 year
- *Opened Canned Foods*: Store tightly covered in a glass or plastic storage container in refrigerator.
  - Baby food: 2 days
  - Meat, seafood, poultry: 2 days
  - Vegetables: 3 days
  - Tomato sauce, fruits: 5 to 7 days
  - Pickles and olives: 1 to 2 months

Cooking of Canned Goods

Commercially canned foods can be safely eaten straight from the can as long as the container is intact. However, Do Not use home canned vegetables unless you have the means to boil them for 10 minutes before eating. Don't taste or use canned foods that show any signs of spoilage! Look closely at all cans before opening them. A bulging lid or leaking can is a sign of spoilage.

When you open the can, look for other signs such as spurting liquid, an off odour or mold. Spoiled canned foods should be discarded so they will not be eaten by humans or pets. Remember that once a can is opened, it becomes perishable and should be refrigerator-stored or cooked properly and then stored in the refrigerator if you are not going to eat it right away.

Serving Canned Goods

Serve foods on clean plates, using clean utensils. Keep

opened canned foods in the refrigerator until it is time to serve them. Keep cooked canned foods hot, above 140 °F, but not for more than two hours. When outside temperatures are 90 °F or warmer, hold the cooked foods no longer than one hour before reheating, refrigerating or freezing them.

Leftovers

After a meal, refrigerate or freeze uneaten food promptly, but throw it out if the food has been held too long or was a reheated leftover at this meal. Refrigerate or freeze cooked leftovers in small, covered shallow containers within two hours after cooking. Leave airspace around the containers for quick chilling. Date packages of leftovers and use them within a safe period. Cover and reheat leftovers thoroughly before serving them.

Bring sauces, soups and gravies to a rolling boil; reheat all other foods to an internal temperature of 165 °F. Reheat only the amount of food needed for one meal. Never reheat a food more than once. Don't taste leftovers to determine if they are safe to eat. If in doubt, throw it out! Discard outdated, unsafe or questionable leftovers. Place them in the garbage disposal or in tightly wrapped packages that cannot be consumed by people or animals.

Handling of Cheese for Safety and Quality

Cheese comes in many forms ranging from soft to hard and natural to processed. How cheese is handled for safety and best quality is dependent on the type of cheese. Natural cheese can be unripened or ripened. Unripened cheeses include cream cheese and cottage cheese. Cheeses that are ripened by bacteria include cheddar, Swiss and Parmesan. Cheeses that are ripened by mold include Blue, Roquefort and Brie.

Natural cheeses are also categorized by their degree of hardness:

- *Soft:* Brie, Camembert, cottage cheese, cream cheese, ricotta, feta
- *Semi-soft:* Blue, brick, Havarti, Monterey Jack, mozzarella, Muenster, provolone
- *Hard:* Cheddar, Colby, Edam, Gouda, Swiss
- *Very Hard:* Parmesan, Romano

Process cheese is a blend of fresh and aged natural cheese that has been melted, pasteurized and mixed with an emulsifier. Process cheese is milder in flavour and softer than the natural cheese from which it was made. It also has very good melting properties.

How should Cheese be Stored

Refrigerate all cheese between 35 and 40 °F in its original wrapping until ready to use. It is best to store cheese in a refrigerator drawer so it does not pick up off-flavours from other foods. If cheese is removed from its original packaging, wrap it tightly with plastic film wrap or foil to prevent air pockets.

Once cheese is exposed to air, molding and dehydration might occur. To protect cheese from mold it is best to double-wrap cheese and place it in a sealed container after each use. Strong-smelling cheeses, like Limburger, should be well-wrapped and kept in a separate container to prevent odours from transferring to other foods.

How Long is it Safe to Keep Cheese

The "best if used by" date tells you how long the product will keep its best flavour or quality. The term is not a safety date. "Best if used by" dates are general guidelines, as some foods may deteriorate more quickly and other foods may last longer than the times suggested.

A number of factors, such as improper handling and inadequate storage, can shorten the shelf-life of a food. Many foods can be eaten after the "best if used by" date if properly stored and handled.

Caution

Throw out soft cheeses that have been at room temperature for more than four hours.

Natural Cheese

As a general rule, the harder the cheese, the longer its shelf-life. But remember, many types of natural cheese will continue to ripen, no matter how carefully they are stored. Hard cheeses will generally keep for several months, whereas softer cheeses will keep from one to three weeks after opening. Large pieces of cheese tend to keep longer than shredded cheese.

Process Cheese

Unopened packages of process cheese can be used up to six months if refrigerated at 40 °F or below. Opened packages should be tightly rewrapped, refrigerated, and used within three to four weeks.

Does Cheese Freeze Well

Most hard cheeses and process cheeses can be frozen, but there will be changes in texture. For this reason, thawed cheese is best used crumbled or shredded, in salads or as toppings or in cooked dishes.

Tips for Freezing Cheeses:

- Freeze pieces of a half-pound or less.
- Use moisture-proof and airtight wrapping.
- Freeze quickly and store at 0 °F for two to six months.
- Thaw in refrigerator so cheese won't lose moisture; the slower the cheese is thawed, the better.
- Use as soon as possible after thawing

Should Moldy Cheese be Thrown away

Molds spread through cheese by producing "hyphae," which are root-like structures. In hard cheeses, such as cheddar and many processed cheeses, these mold hyphae do not penetrate very far below the surface. Mold that is not part of the manufacturing process of a cheese should be handled.

Hard Cheese

If the block of cheese is large and molding is not extensive, remove the mold to save the remaining cheese. Remove at least 1-inch around and below the mold spot. Keep the knife out of the mold itself so it will not contaminate other parts of the cheese. Mold requires oxygen to grow, so tightly rewrap the remaining cheese. The old advice to wipe the cheese with a vinegar solution is no longer recommended and is not required to prevent further mold growth.

Semi-Soft and Soft Cheese and All Types of Crumbled, Shredded, and Sliced Cheese

Discard soft cheese, cheese that has a loose-knit curd, and

shredded or sliced cheese if it contains molds that are not a part of the manufacturing process. These molds can be dangerous. Foods with high moisture content can be contaminated below the surface. Shredded, sliced, or crumbled cheese can be contaminated by the cutting instrument. Moldy soft cheese can also have bacteria growing along with the mold.

- *Caution*: Never open extremely moldy cheese in the kitchen. Do not sniff the moldy item. Wrap all moldy trimmings and moldy cheeses in plastic wrap and place in a covered trash can.

Who should Avoid Eating Certain Cheeses

Pregnant women and newborns, older adults, and people with weak immune systems due to cancer treatments, AIDS, diabetes or kidney disease are at risk for becoming seriously ill from eating foods that contain Listeria monocytogenes. People in these at-risk groups should not eat feta, Brie, Camembert, blue-veined cheeses, or Mexican-style soft cheeses such as queso blanco and queso fresco unless labeled pasteurized.

What is the Best Way to Melt Cheese

Cheese cut into small pieces or shredded promotes more even melting in a shorter amount of time. When you add cheese to any recipe, cook on low heat, stirring constantly. High heat will toughen cheese and make it stringy. When you are making a sauce with cheese in it, add cheese as the last ingredient and heat until just melted. Process cheese melts more smoothly than natural cheese.

What is the Best Way to Microwave Cheese

Remove the wrapping and place on a microwave-safe plate. Microwave at 30 per cent until cheese reaches desired softness and/or temperature. Check every 10 seconds to prevent overheating.

Cooking times will vary among microwave ovens. Use this method to prepare cheese nachos without making the cheese tough.

What is the Best Temperature for Serving Cheese

The flavour of cheese is best when eaten at room temperature, so remove from refrigerator before serving time. Soft cheeses take a shorter amount of time to come to room temperature than firm and hard cheeses. Only set out the amount of cheese you will eat to prevent the cheese from becoming dry and tough from being repeatedly warmed and chilled.

What is the Best Way to Shred Cheese

Cheese will shred more easily if well-chilled. It can also be placed in the freezer for 30 minutes before shredding.

Safe Handling of Eggs

Shell Egg Safety

Eggs can be a part of a healthy diet. However, they are perishable just like raw meat, poultry and fish. To be safe, they must be properly stored, handled and cooked.

Concern for Egg Safety

Some unbroken fresh shell eggs may contain Salmonella enteritidis bacteria that can cause food borne illness. While the number of eggs affected is quite small, there have been some scattered outbreaks in the last few years. Currently the government, the egg industry and the scientific community are working together to solve the problem.

Researchers say that if present, the Salmonella enteritidus bacteria are usually in the yolk or "yellow." But they cannot rule out the bacteria being in egg whites. So everyone is advised against eating raw or undercooked egg yolks, whites or products that contain eggs.

People with health problems, the very young, the elderly and pregnant women are particularly vulnerable to Salmonella enteritidis infections. Chronic illness also weakens the immune system, making the person vulnerable to food borne illness.

Storage, Handling and Cooking of Shell Eggs

Proper refrigeration, cooking and handling should prevent most egg safety problems. People can enjoy eggs and dishes containing eggs if these safe handling guidelines are followed.

Don't Eat Raw Eggs

This includes "health food" milk shakes with raw eggs, Caesar salad, hollandaise sauce and any other foods like homemade mayonnaise, ice cream or eggnog made from recipes in which the raw egg ingredients are not cooked. These egg-based recipes should be updated to start with a cooked base or so that commercially prepared pasteurized eggs or egg substitutes are used. Use a thermometer and make sure the temperature of the cooked base reaches 160 °F.

Buy Clean Eggs from a Refrigerator Display Case

Do not purchase eggs anywhere that are not refrigerated. Any bacteria present in the egg can grow quickly if stored at room temperature. At the store, choose Grade A or AA eggs with clean, uncracked shells.

Safe Storage of Eggs at Home

Take eggs straight home and store them immediately in the refrigerator at 40 °F or slightly below. Store them in the grocery carton in the coldest part of the refrigerator, not in the door. Do not wash eggs. Washing eggs could remove the protective mineral oil coating put on at the plant and could increase the potential for bacteria on the shell to enter the egg.

Use Eggs Promptly

Use raw shell eggs within three to five weeks. When fresh eggs are hard cooked, the protective coating is washed away so hard-cooked eggs should be refrigerated within two hours of cooking and used within a week. Use leftover yolks and whites within four days. If eggs crack on the way home from the store, break them into a clean container, cover tightly, and keep refrigerated for use within two days.

Freeze Eggs for Longer Storage

Eggs should not be frozen in their shells. To freeze whole eggs, beat yolks and whites together. Egg whites and yolks can also be frozen by themselves. Use frozen eggs within a year. If eggs freeze accidentally in their shells, keep them frozen until needed. Defrost them in refrigerator. Discard any with cracked shells.

Handle Eggs Safely

Wash hands, utensils, equipment and work areas with warm, soapy water before and after contact with eggs and egg-rich foods.

Serve Immediately

Don't keep out of the refrigerator more than two hours. Serve cooked eggs and egg-rich foods immediately after cooking, or place in shallow containers for quick cooling and refrigerate at once. Use within three to four days. Recipes using raw eggs should be cooked immediately or refrigerated and cooked within 24 hours.

Cooked Eggs

Hard-cooked eggs should be safe for everyone to eat. Those at risk for food borne illness should avoid eating soft-cooked or "runny" eggs. However, healthy persons may choose to eat eggs that are less than totally firm.

Use the following cooking times:

- *Fried Eggs:* Cook 2 to 3 minutes on each side; 4 minutes in a covered pan. The yolk should begin to thicken.
- *Scrambled Eggs*: Cook until firm throughout.
- *Poached Eggs*: Cook 5 minutes over boiling water.
- *Soft-Cooked Eggs*: Cook 7 minutes in the shell in boiling water.

Use Safe Egg Recipes

Egg mixtures are safe if they reach 160 °F, so homemade ice cream and eggnog can be made safely from a cooked base. Heat the egg-milk mixture gently. Use a thermometer or be sure the mixture coats a metal spoon. Dry meringue shells are safe. So are divinity candy and seven-minute frosting, made by combining hot sugar syrup with beaten egg whites. Meringue-topped pies should be safe if baked at 350 °F for about 15 minutes.

Chiffon pies and fruit whips made with raw, beaten egg whites cannot be guaranteed safe. Substitute whipped cream or whipped topping. To make key lime pie safely, heat lime juice with raw egg yolks in a pan on the stove, stirring constantly, until the mixture reaches 160 °F. Combine this mixture with sweetened condensed milk and pour it into a baked piecrust. For meringue topping. For egg dishes such as quiche and casseroles, a knife inserted in the center should come out clean.

Safety Tips for Easter Eggs:

- Wash hands before dyeing the eggs.
- Use a food-safe colouring if eggs will be eaten.
- Handle eggs carefully to prevent cracking. If the shells crack, bacteria could contaminate the inside. Do not use cracked eggs for hiding.
- After hard-cooking eggs, dye them and return them to the refrigerator immediately until ready to hide.
- Hide eggs in places protected from dirt, pets and other sources of bacteria.
- The total time for hiding and hunting eggs should not exceed two hours. Discard any Easter eggs that have been cracked. The uncracked "found" eggs should then be refrigerated until eaten.
- Eat hard-cooked eggs within one week.

Ukranian Easter Eggs

A traditional part of the Easter celebration in the Ukraine, these whole, raw eggs have their contents blown out to leave a hollow shell that is then decorated. Because some raw eggs may contain Salmonella, use caution when blowing out the contents.

Use only uncracked eggs that have been kept refrigerated. To destroy bacteria that may be present on the surface of the egg, wash the egg in hot water and then rinse in a solution of one teaspoon chlorine bleach per half cup of water. After blowing out the egg, refrigerate the contents and use within two to four days; cook thoroughly before eating.

Pickled Eggs

Pickled or brined eggs are hard-cooked eggs marinated in vinegar and pickling spices, spicy cider, or juice from pickles or pickled beets. Home-pickled eggs should be kept refrigerated. Unopened containers of commercially processed pickled eggs keep for several months on the shelf. After opening, keep refrigerated.

Egg Products

When consumers think of an egg, most picture a pristine white, or sometimes brown, oval shell containing a completely edible, versatile food. But of the 63 billion eggs consumed in 1994, more than 25 per cent were in the form of egg products. The term "egg products" refers to eggs that have been removed from their shells for processing. Basic egg products include whole eggs, whites, yolks and various blends with or without non-egg ingredients that are processed and pasteurized and may be available in liquid, frozen and dried forms. Egg products are widely used by the food service industry and as ingredients in other foods, such as prepared mayonnaise and ice cream.

Buying, Handling and Cooking Egg Products

Proper storage and handling is necessary for all egg products to prevent bacterial contamination:

- Buy only pasteurized egg products that bear the USDA inspection mark.
- Make sure containers are tightly sealed. Frozen products should show no signs of thawing. Refrig-erated products should be kept at 40 °F or below. Dried egg products should not be caked or hardened.
- Store frozen egg products up to one year at 0 °F or lower. After thawing, do not re-freeze.
- Thaw frozen egg products in the refrigerator or under cold running water. Do Not Thaw On The Counter.
- If the container for liquid products bears a "use by" date, observe it. Follow the storage and handling instructions provided by the manufacturer.
- For liquid products without an expiration date, store unopened containers at 40 °F or below for up to seven days. Do not freeze opened cartons of liquid egg products.
- Unopened dried egg products and egg white solids can be stored at room temperature as long as they are kept cool and dry. After opening, store in the refrigerator.
- Reconstituted egg products must be used immediately or refrigerated and used that day.
- USDA Commodity Dried Egg Mix should be stored at less than 50 °F, preferably in the refrigerator. After

opening, use within seven to 10 days. Reconstitute only the amount needed at one time and use immediately or refrigerate and use within an hour.

Egg Product Safety

Who Inspects Egg Products

Congress passed the Egg Products Inspection Act (EPIA) in 1970. The EPIA provides for the mandatory continuous inspection of the processing of liquid, frozen and dried egg products to ensure they are wholesome and properly labeled and packaged to protect the health and welfare of consumers. USDA's Food Safety and Inspection Service (FSIS) inspects all egg products, with the exception of those products exempted under the Act that are used by food manufacturers, food service institutions and retail markets.

The Department of Health and Human Services' Food and Drug Administration (FDA) is responsible for the inspection of egg substitutes, imitation eggs and similar products that are exempted from continuous inspection under the EPIA. Currently, only Canada is exporting egg products to the United States. The EPIA specifies that egg products may not be imported into the United States except from countries that have an egg products inspection system equivalent to that in this country.

Why are Eggs Products Useful

Food manufacturers like pasteurized egg products because of their convenience and ease in handling and storing. Institutional food service operators, such as fast food chains, restaurants, hospitals and nursing homes, use egg products to ensure a high level of food safety. Consumers are starting to see more egg products in retail food stores.

How are Egg Products Made

Egg products are processed in sanitary facilities under continuous inspection by the USDA. The initial step in making egg products is breaking the eggs and separating the yolks and whites from the shells. Eggs are processed by automated equipment that moves the eggs from flats, washes and sanitizes the shells, breaks the eggs and separates the whites and yolks, and/or makes mixtures of them. The liquid egg product is filtered, mixed, and then chilled prior to additional processing.

Why and how are Egg Products Pasteurized

The law requires that all egg products distributed for consumption be pasteurized. This means that they must be rapidly heated and held at a minimum required temperature for a specified time. This destroys Salmonella, but it does not cook the eggs or affect their colour, flavour, nutritional value or use. Dried whites are pasteurized by heating in the dried form, again for a specified time and at a minimum required temperature. Since many new and different types of egg products are now being formulated, government and industry researchers are currently evaluating the effectiveness of the pasteurization processes used for these and other products. Additional research will determine if new safety measures are needed to continue to provide safe egg products for food service, industry and consumers.

Are All Egg Products Pasteurized

Certain commodities are not presently considered egg products and are exempt from this law. These commodities include freeze-dried products, imitation egg products and egg substitutes. Inspected, pasteurized egg products are used to make these commodities and companies may elect to repasteurize these products following formulation and before packaging. Officially inspected egg products will bear the USDA inspection mark. No-cholesterol frozen egg substitutes first became available to consumers in 1973. They consist of egg whites, artificial colour and other non-egg additives. Specific questions about egg substitutes should be directed to the manufacturer.

What Information is Provided on Egg Product Labels

In addition to nutrition information on consumer packages, other labeling information is required for egg products.

All egg products must be labeled with:

- The common or usual name;
- The ingredients listed in the order of descending proportions;

- The name and address of the packer or distributor;
- The date of pack which may be shown as a lot number or production code number;
- The net contents; and
- The official USDA inspection mark and plant number.

Can Egg Products be Used as an Ingredient in Uncooked Foods

Egg products are pasteurized and can be substituted in recipes typically made with raw eggs that won't be cooked. For optimal safety, egg products are best used in a cooked product, especially if serving high-risk persons such as the elderly, children, the chronically ill, or those with weakened immune systems. Use a food thermometer to be sure that the internal temperature of the cooked product reaches 160 °F. Egg products can also be used in baking or cooking.

What is Dried Egg Mix

USDA Dried Egg Mix is a blend of dried whole eggs, nonfat dry milk, soybean oil and a small amount of salt. There is very little moisture in it. To reconstitute, blend 2 Tablespoons of Dried Egg Mix with ¼ cup water to make the equivalent of one large whole egg. Dried Egg Mix is packaged in 6-ounce pouches, equivalent to about six eggs each.

It is distributed by USDA to food banks, Indian reservations and other needy family outlets and is also used in disaster feeding. A similar product called All-Purpose Egg Mix, containing a greater proportion of eggs, is now being manufactured for USDA. It is reconstituted by mixing one part egg mix with two parts water. All-Purpose Egg Mix is available to schools as part of the School Lunch Programme. It is packaged in 10-pound bags.

Safe Handling of Fish

Selecting the Best

Spotting a Safe Seafood Seller

Always purchase fish from a dealer that maintains high quality.

Based on FDA's Food Code, here are some ways of spotting a safe fish dealer:

- Employees should be in clean clothing and wearing hair coverings.
- They should not be smoking, eating or playing with their hair.
- They should not be sick or have any open wounds.
- Employees should be wearing disposable gloves when handling food and change gloves after doing nonfood tasks and after handling raw fish.
- Fish should be displayed on a thick bed of fresh—not melting—ice preferably in a case or under some type of cover. Fish should be arranged with the bellies down so that the melting ice drains away from the fish, thus reducing the chances of spoilage.

Selecting Quality Fish

Appearance is bright and shiny in quality fish, and most of the scales are intact and adhere tightly to the skin. Each species has characteristic markings and colours which fade and become less pronounced as the fish loses freshness. Eyes are bright, clear, full and often protrude. As quality goes down, the eyes often turn pink and become cloudy and sunken. This does not always apply to small-eyed fish such as salmon. Gills are red and free from slime.

With time, the colour fades to light pink, then gray and finally greenish or dull brown. Odour is fresh and mild. A fish just out of the water has practically no "fish" odour. The fishy odour develops with time, but should not be strong or objectionable. Flesh is firm, elastic and not separating from the bones.

Fresh Fillets and Steaks

Odour is fresh and mild. Flesh is moist, firm, elastic and has a fresh-cut appearance without traces of browning or drying around the edges. Pre-packaged steaks and fillets are tightly wrapped with no liquid and little or no air in the packages.

Safe Handling after Purchase

Whether you've purchased fish that is fresh or frozen,

always keep it cold. Never leave perishable items in a hot car unless packed in ice or in a cooler; seafood products must be kept cold to ensure peak quality. It's always a good idea to keep your refrigerator temperature between 32 to 38 °F, and your freezer at 0 °F or colder.

Storage

Refrigeration

Store fresh fish in its original wrapper in the coldest part of the refrigerator, which is under the freezer or in the "meat-keeper" drawer. Plan to use your fish purchases within one to two days. If not, freeze them. However, do not refreeze previously frozen products because the quality will suffer.

Freezing

It is best to freeze fish in tightly wrapped package form. This takes less storage space and fits a family portion for one meal at a time. Fish freeze faster in tightly wrapped packages. Small whole fish, steaks or fillets are easy to prepare for packaged freezing. Prewrap them tightly and individually in cling wrap, forming a tight skin on the product. Master-bag these individually wrapped items in a good, strong polyethylene bag or foil before freezing, but never more than a pound per master bag.

Large fish have large surface areas exposed, and they are difficult to protect from oxidation. The best way to handle these fish is simply to freeze them unwrapped or temporarily bag them in plastic. After freezing, dip them in water to form a protective glaze.

Then you can rebag the fish and return it to the freezer. The glaze may need renewing every five to six weeks. Label each package with the date, kind and type of fish and the weight and number of servings or pieces. A crayon or grease pencil is ideal for this purpose.

Do not overload your freezer, and do not pack the unfrozen fish too tightly. Either of these practices can greatly extend the freezing time and reduce the quality. Most home-frozen fish should not be stored over six months, no more than three months for salmon.

Preparation

Cleanliness

Always wash hands thoroughly with hot, soapy water before preparing foods and after handling raw fish. Don't let raw fish or juices touch ready-to-go foods either in the refrigerator or during preparation. Don't put cooked foods on the same plate that held raw fish. Always wash utensils that have touched raw fish with hot, soapy water before using them for cooked foods. Wash counters, cutting boards and other surfaces raw fish have touched.

Thawing

While freezing fish quickly keeps more cell walls intact, the opposite is true for thawing. Defrost gradually so cells are disturbed less and fewer juices leak out. The best way to thaw is overnight in the refrigerator. Avoid thawing at room temperature. If you must thaw fish quickly, here are safe options: seal fish in a plastic bag and immerse in cold water for about an hour, or microwave on the "defrost" setting, stopping when fish is still icy but pliable.

Marinating

Marinate fish in the refrigerator, not on the counter. Discard the marinade after use because it contains raw juices, which may harbor bacteria. If you want to use the marinade as a dip or sauce, reserve a portion before adding raw food.

Cooking

Guidelines for Cooking Fish

Cooked to perfection, fish is at its flavourful best and will be moist, tender and have a delicate flavour. In general, fish is cooked when its meat just begins to flake easily when tested with a fork and it loses its translucent or raw appearance. Like most foods, fish should be thoroughly cooked. The U.S. Food and Drug Administration (FDA) suggests cooking fish until it reaches an internal temperature of 145 °F. One helpful guideline is the 10-minute rule for cooking fish. Apply it when baking, broiling, grilling, steaming and poaching fillets, steaks or whole fish. Practice makes perfect and cooking fish properly is all in the timing.

Here's how to use the 10-minute rule:

- Measure the seafood product at its thickest point. If the fish is stuffed or rolled, measure it after stuffing or rolling.
- At 450 °F bake for 10 minutes per inch thickness of the fish, turning the fish halfway through the cooking time. For example, a 1-inch fish steak should be cooked 5 minutes on each side for a total of 10 minutes. Pieces of fish less than half an inch thick do not have to be turned over.
- Add 5 minutes to the total cooking time if you are cooking the fish in foil or if the fish is cooked in a sauce.
- Double the cooking time for frozen fish that has not been defrosted.

Fish is the original "fast food." It cooks quickly, within minutes, because it lacks the connective tissue of red meats and poultry. Some of the best cooking methods for fish include poaching, broiling, grilling, baking and microwaving because they bring out flavour without adding fat.

Baking

Whole fish, whole stuffed fish, fillets, stuffed fillets, steaks and chunks of fish may be baked. Use pieces of similar size for even cooking. It's best to bake fish in a preheated, 450 °F oven following the 10-minute rule; bake uncovered, basting if desired.

Broiling

Steaks, whole fish, split whole fish and fillets lend themselves well to broiling. Place fish, 1-inch thick or less, 2 to 4 inches from the heat source. Place thicker pieces 5 to 6 inches away.

Baste frequently with an oil-based marinade. Using the 10-minute rule, cook on one side for half the total cooking time, basting once or twice, then turn the fish over to continue broiling and basting.

Grilling

This technique lends itself well to thick steak fish such as salmon, halibut, swordfish, tuna and whole fish. Preheat an outdoor gas or electric grill. If using a barbecue grill, start the fire about 30 minutes before cooking. Let it burn until white-hot then spread coals out in a single layer. Adjust the grill height to 4 to 6 inches above the heat.

To grill fish, a moderately hot fire is best for cooking seafood. Always start with a well-oiled grid to prevent the delicate skin of the fish from sticking. Support more delicate pieces of fish in a hinged, fish-shaped wire basket for easier turning or handling. Frequently baste steaks and fillets while grilling to prevent them from drying out.

Marinating fish an hour before grilling also helps keep it moist. Apply the 10-minute rule for proper doneness. Use indirect heat for whole fish by banking hot coals on either side of the barbecue or preheat gas or electric grill. Oil fish well and place in an oiled fish basket. Cook fish covered, 10 to 12 minutes per inch of thickness, turning halfway through cooking time.

Microwaving

Use a shallow dish to allow maximum exposure to the microwaves. Arrange fillets with the thicker parts pointing outward and the thinner parts, separated by pieces of plastic wrap, overlapping in the center of the dish. Cover dish with plastic wrap and vent by turning back one corner. Allow 3 minutes per pound of boneless fish cooked on high as a guide. Rotate the dish halfway through the cooking time. Rolled fillets microwave more evenly and are less likely to overcook than flat fillets, which may have thin edges.

Safe Handling of Lamb

Selecting the Best

Fresh is Best

Select lamb that is pinkish red with a velvety texture. Dark red cuts generally indicate the meat is older and less tender.

Look for good marbling, and meat that is fine textured and firm:

- Look for packages that are cool to the touch and have no wear or punctures.

- Always remember to select meat just before checking out at the supermarket register.
- Make sure all meats, whether raw, pre-packaged or from the deli are refrigerated when purchased.
- Prevent "cross-contamination" by keeping fresh meats separate from other items. Put raw meat packages in a plastic bag so juices won't drip onto other foods.
- Pack raw meats in an ice chest if it will take more than an hour to get home. Keep ice chest in the passenger area of the car during warm weather.
- Take meats straight home to the refrigerator or freezer.

Product Dating

Product dating, applying "sell by" or "use by" dates, is not required by federal regulations. However, many stores and processors may voluntarily choose to date packages of raw lamb. Use or freeze products with a "sell by" date within three to five days of purchase.

If the manufacturer has determined a "use by" date, observe it. It's always best to buy a product before its date expires. It's not important if a date expires after freezing lamb because all foods stay safe while properly frozen.

Product Inspection and Grading

All lamb found in retail stores is either USDA-inspected for wholesomeness or inspected by state systems which have standards equal to the federal government. Each animal and its internal organs are inspected for signs of disease.

The "Passed and Inspected by USDA" seal ensures that the lamb is wholesome and free from disease. Although inspection is mandatory, its grading for quality is voluntary, and a plant pays to have its lamb graded.

USDA Lamb Quality Grades are based upon palatability-indicating characteristics of the lean and carcass conformation. Conformation has no direct influence upon the eating quality. For lamb and yearling mutton the quality grades are Prime, Choice, Good and Utility.

Table. Safe Handling of Lamb in Cold Storage.

Food	Refrigerator (40 °F)	Freezer (0 °F)
Fresh Lamb		
Roast	3-5 Days	6-9 Months
Steak/Chops	3-5 Days	6-9 Months
Ground/Stew Meat	1-2 Days	3-4 Months
Cooked Lamb		
Roast	3-4 Days	2-3 Months
Steak/Chops	3-4 Days	2-3 Months
Ground/Stew Meat	1-2 Days	2-3 Months

Storing

Refrigeration

Packaged whole cuts of fresh lamb may be refrigerated in their original wrapping in the coldest part of the refrigerator up to four or five days after purchase. Ground lamb can be stored in the refrigerator for up to two days; cooked lamb is at its best refrigerated no longer than four days.

- Use an appliance thermometer and maintain a temperature of 34 to 38 °F.
- Never leave meat in a hot car or sitting out at room temperature. Refrigerate or freeze fresh lamb Immediately afer bringing it home.
- Store uncooked lamb items together, separate from cooked foods. Make sure the raw juices do not drip onto other foods.
- Keep lamb refrigerated until you are ready to cook it.
- When transporting uncooked or cooked lamb to another dining site, place it in an insulated container or ice chest until ready to cook or eat.

Freezing

Freeze whole cuts of fresh lamb if you do not plan to cook it within four days after purchase. Wrap whole cuts of lamb separately in heavy-duty foil or moisture-and vapour-proof freezer bags or wrap before freezing. Label for ease in selecting just the right number of cuts to thaw for a single meal. Be sure to press the air out of the package before freezing. If you plan to freeze lamb in its original wrapping, overwrap the porous store plastic with freezer bag or paper. Cooked lamb cuts may be frozen in the same way as fresh, unless made with a sauce or gravy. In that case, pack the meat in a rigid container with a tight-fitting lid.

Preparation

Cleanliness

- Always wash hands thoroughly with hot, soapy water before preparing foods and after handling raw meat.
- Don't let raw meat or juices touch ready-to-eat foods either in the refrigerator or during preparation.
- Don't put cooked foods on the same plate that held raw lamb.
- Always wash utensils that have touched raw meat with hot, soapy water before using them for cooked meats.
- Wash counters, cutting boards and other surfaces raw meats have touched.

Thawing

Thaw uncooked lamb in the refrigerator or in cold water. Never thaw meat at room temperature. Thawing by refrigeration requires planning ahead, most likely allowing a 24-hour thawing period. After defrosting raw lamb by this method, it will be safe in the refrigerator up to five days before cooking.

During this time, if you decide not to use the lamb, you can safely refreeze it without cooking it first. To thaw lamb in cold water, leave the meat in its original wrapping or place it in a watertight plastic bag. Change the water every 30 minutes. For quick thawing of uncooked or cooked lamb, use the microwave, but plan on cooking the meat immediately after thawing because some areas of the food may become warm and begin to cook during microwaving.

Thawing time will vary according to whether you're thawing a whole roast or cuts and the number of parts frozen together. Use the Defrost or Medium-Low setting, according to the manufacturer's directions. Turn the roast and separate parts as they thaw, taking care the meat does not begin to cook. Repeat as needed. Foods defrosted by the cold water method or in the microwave should be cooked before refreezing because they may have been held at temperatures above 40 °F.

Marinating

Marinate food in the refrigerator, not on the counter. Discard the marinade after use because it contains raw juices, which may harbor bacteria. If you want to use the marinade as a dip or sauce, reserve a portion before adding raw food.

Partial Cooking or Browning

Never brown or partially cook lamb, then refrigerate and finish cooking later, because any bacteria present would not have been destroyed. It is safe to partially pre-cook or microwave lamb immediately before transferring it to the hot grill or oven to finish cooking.

Cooking

Importance of Kitchen Thermometers

One of the critical factors in controlling bacteria in food is controlling temperature. Pathogenic microorganisms grow very slowly at low temperatures, multiply rapidly in mid-range temperatures and are killed at high temperatures. For safety, foods must be cooked thoroughly.

It is essential to use a thermometer when cooking meat and poultry to prevent undercooking and, consequently, prevent food borne illness. Using a thermometer is the only reliable way to ensure safety and to determine the "doneness" of most foods. To be safe, a product must be cooked to an internal temperature high enough to destroy any harmful bacteria that may have been in the food. Recent research has shown that colour and texture indicators are not reliable.

Cooking the Meat

For safety, the USDA recommends cooking ground lamb patties and other ground mixtures to 160 °F. Whole muscle meats such as chops and roasts should be cooked to 145 °F, 160 °F or 170 °F. Remember that appliances and outdoor grills can vary in heat. Use a meat thermometer to check for safe cooking and doneness of lamb. Cooked muscle meats can be pink even when the meat has reached a safe internal temperature. If fresh lamb has reached 160 °F throughout, even though it may still be pink in the center, it should be safe. The pink colour can be due to the cooking method or added ingredients.

Microwaving

When microwaving unequal sizes of lamb, arrange in a dish or on a rack so thick parts are towards the outside of the dish and thin parts are in the center, and cook on medium-high or medium power. Place a roast in an oven-cooking bag or in a covered pot. Refer to the manufacturer's directions that accompany the microwave oven for suggested cooking times. Use a meat thermometer to test for doneness in several places to be sure the proper temperature.

Serving

- Wash hands with soap and water before serving or eating food.
- Serve cooked products on clean plates with clean utensils and clean hands.
- Never put cooked foods on a dish that has held raw lamb unless the dish is washed with soap and hot water.
- Hold hot foods above 140 °F and cold foods below 40 °F.
- Never leave foods, raw or cooked, at room temperature longer than two hours. On a hot day with temperatures at 90 °F or warmer, this decreases to one hour.

Safe Handling of Milk and Dairy Products

Milk

Grade A milk is carefully produced, processed and packaged in order to protect the safety of the consumer. Grade A milk must be pasteurized to be sold by retailers in interstate commerce. Raw milk is usually pasteurized either by low temperature pasteurization in which the milk is heated to 145 °F or higher for at least 30 minutes, or by high temperature pasteurization in which the milk is heated to 161 °F or higher for at least 15 seconds and then quickly cooled. Pasteurization destroys disease-causing bacteria and extends the shelf life of milk. However, pasteurized milk can readily spoil and could cause food borne illness if not properly protected and handled.

Maintaining the Safety of Milk

Refrigeration is the single most important factor in maintaining the safety of milk. By law, Grade A milk must be maintained at a temperature of 45 °F or below. Bacteria in milk will grow minimally below 45 °F. However, temperatures well below 40 °F are necessary to protect the milk's quality. It is critical that these temperatures be maintained through warehousing, distribution, delivery and storage.

The cooler refrigerated milk is kept, the longer it lasts and the safer it is. As the product is allowed to warm, the bacteria grow more rapidly. Properly refrigerated, milk can withstand about two weeks' storage. Infants, pregnant women, the elderly and the chronically ill are most at risk from serious illness due to eating any unsafe food. These individuals and those who care for them must be especially careful to handle milk safely.

Fresh Fluid Milk

Fresh milk is categorized mainly by the amount of butterfat it contains. In November 1997, the FDA announced a new rule for milk labeling that helps consumers clarify the difference between 1-and 2-per cent milk and reinforces the fact that skim milk is fat-free. Under the new rule, 2-per cent milk is renamed reduced fat; 1-per cent milk is renamed low-fat; and skim milk is called fat-free or nonfat, although it may contain up to 0.5 grams of fat in a one cup serving.

Buying Fresh Fluid Milk

When selecting milk at the store, make sure it is properly displayed and pay close attention to the date on the label. All fresh fluid milks should be stored at temperatures below 40 °F and should not be stacked high in the display cases. If stored above 40 °F, milk will begin to develop signs of spoilage, including sour odour, off-flavour and curdled consistency. Remember that milk should be taken from the store and quickly placed in your refrigerator at home so that the temperature does not rise above 40 °F.

Once you have purchased milk and refrigerated it promptly, it should retain its fresh taste for one to five days beyond the "sell-by" date if kept at the proper temperature. If it spoils before the date expires, this indicates it was not handled properly, and it should be returned to the store for a refund.

Storing Fresh Fluid Milk

Milk should not be left out at room temperature. Pour milk to be used into a serving container and return the original container to the refrigerator. Do not return unused milk that has been sitting out to its original container where it could contaminate the remaining milk. Milk can be stored frozen at 0 °F for up to three months and will be safe to drink if it is thawed in the refrigerator, although it does not retain its smooth texture.

Buttermilk

Originally, buttermilk was made as a by-product when making butter. Lactic acid bacteria are added to fresh, fluid pasteurized skim or part-skim milk to produce the thick, tangy buttermilk. Buttermilk should be handled with the same precautions as regular fluid milk.

Flavoured Milk

Chocolate and other flavours-such as maple, strawberry and coffee-may be used for flavoured milks. These milks are stored and used as fresh fluid milk.

Concentrated or Dried Milks

Evaporated and Evaporated Skimmed Milk

This type of milk has about 60 per cent of the water removed. It may be fortified with vitamins A and/or D. Store cans of evaporated milk in a cool, dry place. It is shelf-stable, but once opened it should be treated as fresh fluid milk, kept refrigerated and used within several days.

Sweetened Condensed Milk

This is the milk that results from the evaporation of half the water and the addition of sugar in amounts sufficient for preservation. It is stored like evaporated milk.

UHT Milk

Ultra-high temperature (UHT) milk is regular fluid milk that is packed in an airtight, sterilized, cardboard container. The product is treated by flash sterilization at 290 °F. This high temperature kills all bacteria or microorganisms. The milk is then packed into sterilized containers and is shelf-stable for six months.

After six months, the flavour and colour begin to change and the product thickens. It is still safe, but may not produce the desired effect in a recipe. Once the package of UHT milk is opened, it is treated like fresh fluid milk and used within several days.

Nonfat Dry Milk

This dairy product is made by removing water from pasteurized, fat-free milk. Due to its low moisture content, it can be kept for long periods of time. Once reconstituted, it is handled like fresh fluid milk.

Cream

Basic Facts

Cream has a very high fat content of between 18 and 40 per cent butterfat compared to around 3.25 per cent in whole milk. The rich, yellow colour associated with cream comes from the carotene in the fat.

The type of cream is determined by its fat content:

- Half-and-half is a combination of milk and cream with a butterfat content of about 11 per cent.
- Light cream has between 18 and 30 per cent butterfat and may be called coffee or table cream.
- Light whipping cream has between 30 and 36 per cent butterfat and can be whipped into solid form, although it tends to be less stable than heavy whipping cream.
- Heavy cream or whipping cream contains 36 to 40 per cent butterfat.
- Pressurized whipped cream is sold in aerosol cans and is made from a mixture of cream, sugar and chemical stabilizers.

- Sour cream is made by adding a lactic acid culture to sweet cream. Sour cream usually contains between 18 and 20 per cent butterfat.
- Reduced-fat sour cream has skim milk added to lower the fat content.

Storage

Store cream at 40 °F or below in its original container in the refrigerator. Do not leave cream at room temperature, and do not mix warm cream with cream that has been kept refrigerated. Use fresh, pasteurized cream within one to five days of the "sell-by" date.

Nondairy Dessert Toppings

Nondairy dessert toppings are made from vegetable oils but may also contain some milk products. The frozen toppings may be stored for up to one year in the freezer, or thawed and kept in the refrigerator for up to two weeks. Do not freeze dessert toppings in aerosol cans. Store the cans in the refrigerator for a maximum of two to three months.

Butter

Basic Facts

Butter is made from the sweet or soured cream of cow's milk by agitation or "churning." After churning, the mass of butter is washed and salted. It is worked to distribute the salt and remove extra water. To be sold in stores, butter must contain at least 80 per cent milk fat.

Water and milk solids make up the other 20 per cent. Salt and colouring may be added if desired. Some unsalted butter is sold as sweet butter, but most people prefer the salted product. The USDA grade label on the butter carton or wrapper means the butter has been tested for quality by a government grader and has been produced under sanitary conditions.

The highest possible grade is AA. Grade AA butter is delicate and sweet-flavoured with a creamy texture and good spreadability. Most butter sold is grade AA or A. Because of its high fat content, butter contains many calories and therefore should be used sparingly in the diet.

Storage

Storing butter properly, lengthens the shelf life so it can be used over a longer period of time. To prevent a type of spoilage called rancidity, protect butter from heat, light and air by storing it covered in the refrigerator. Rancid butter has an unpleasant taste and smell.

Butter absorbs odours from other foods rapidly. To prevent flavour changes, keep butter wrapped in moisture-and vapour-proof material or in tightly covered containers. For refrigerator storage, leave butter in its original wrapper. Opened portions of butter should be refrigerated in a covered dish.

Butter can be stored for up to two weeks at refrigerator temperatures. Higher temperatures cause off-flavours and unpleasant odours to develop. Butter should not be stored in the butter keeper on the refrigerator door longer than two days. For ease in spreading, remove butter from the refrigerator 10 to 15 minutes before using it. For holding longer than two weeks, butter should be frozen.

To store butter in the freezer, wrap it in moisture-and vapour-proof freezer packaging material to keep the butter from absorbing odours from other foods and to prevent freezer burn. Butter in its original carton can be overwrapped. Butter in one-pound blocks can be cut into smaller portions, repackaged and frozen for future use. If properly wrapped and held at 0 °F or lower, butter will keep well in the freezer for six to nine months. Thaw butter in the refrigerator.

Yogurt

Basic Facts

The natural sugar in milk is converted to lactic acid by means of a bacterial culture producing the creamy, pleasantly tart yogurt. Yogurt is pasteurized to destroy disease-causing microorganisms. Fat and calorie content will differ depending on whether whole, low-fat or fat-free milk is used and whether fruit or sweeteners are added. The label will give the specific information for each yogurt.

Storage

Yogurt may be kept well covered in the refrigerator for

seven to 10 days past the "sell-by" date. If it is kept longer, it will develop a stronger taste. Freezing yogurt is not recommended because of the variable results in texture.

Frozen Dairy Products

Basic Facts

This category includes ice cream, ice milk, sherbet and frozen yogurt:

- Ice cream has the highest milk fat and milk solids content. The milk fat content in ice cream usually ranges between 10 and 14 per cent, but may be as high as 20 per cent in specialty ice creams.
- Ice milk often has more sugar than ice cream, but its milk fat ranges from 2 to 7 per cent.
- Sherbet also has less milk fat and milk solids than ice cream, but more sugar and usually contains fruit and fruit acid. Milk fat content of sherbet is between 1 and 2 per cent.
- Frozen yogurt is made from cultured milk and has less milk fat than ice cream and less sugar than sherbet.

Storage

When purchasing ice cream and other frozen desserts at the store, make sure they are frozen solid and that the container is not sticky or frosted which indicates it has partially thawed at some point. Request that the ice cream be placed in an insulated bag or be double bagged to reduce melting on the way home. Ice cream may be stored unopened for up to two months at 0 °F or below.

However, if it will be stored longer than one month, it is best to overwrap the original container with freezer paper or wrap. Once the container has been opened, place plastic wrap over the surface of the ice cream to minimize the development of large ice crystals and the loss of its creamy texture.

Use ice cream within seven to 10 days for best quality. Each time the ice cream is removed from the freezer, and the surface begins to thaw, the ice cream loses quality. If ice cream or other frozen dairy products thaw completely, they should be discarded because of the danger of bacterial growth.

Table. Safe Cold Storage Times for Milk and Dairy Products.

Product	How to Store	Refrigerator (35-40 °F)	Freezer (0 °F)
Pasteurized Fresh Whole or Skimmed Milk	Refrigerate immediately in original container. Keep container closed.	1 to 5 days beyond "sell-by" date	3 months. Freezing may result in change in texture. Thaw in refrigerator.
Sweetened Condensed Milk(Opened)	Refrigerate tightly covered.	1 week	Do not freeze.
Evaporated Milk (Opened)	Refrigerate tightly covered.	1 week	Do not freeze.
Cultured Buttermilk	Refrigerate immediately in original container. Keep container closed.	2 weeks	Do not freeze.
Homogenized, Reconstituted Dry Nonfat and Skimmed Milk	Keep containers tightly closed. Don't return unused milk to original containers.	1 week	Do not freeze.
Sweet and Regular Cream	Refrigerate immediately in original container. Keep container closed.	1 to 5 days beyond "sell-by" date	Do not freeze. (Change of texture, body appearance. Separation of fat emulsion.)
Non-Dairy Whipped Topping	Keep covered.	3 months in aerosol can. 3 days if prepared from mix. 2 weeks if bought frozen and then thawed.	Do not freeze aerosol cans; others may be stored in freezer up to one year.
Butter	Refrigerate immediately in original container. Keep container closed.	2 weeks	Butter made from pasteurized cream: 6 to 9 months.
Sour Cream	Refrigerate immediately in original container. Keep container closed.	2 weeks	Do not freeze.
Ice Cream	Store in original container in freezer.	Do not store here.	2-3 weeks (Opened) 2 Months (Unopened)
Yogurt	Keep covered.	7-10 days	Do not freeze.
Soft Custards, Milk Puddings, Cream and	Cool cooked dishes quickly and refrigerate	5-6 days	Do not freeze.

Safe Handling of Pork

Selecting the Best

Fresh is Best

When buying pork, look for cuts with a relatively small amount of fat over the outside and with meat that is firm and a grayish-pink colour.

For best flavour and tenderness, meat should have a small amount of marbling. Look for packages that are cool to the touch and have no wear or punctures. Always remember to select meat just before checking out at the supermarket register.

Make sure all meats, whether raw, pre-packaged or from the deli are refrigerated when purchased. Fresh meats may contaminate other grocery items. The best way to prevent this "cross-contamination" is to always keep fresh meats separate from other items.

Put raw meat packages in a plastic bag so juices won't drip onto other foods. Pack raw meats in an ice chest if it will take more than an hour to get home. Keep ice chest in the passenger area of the car during warm weather. Take meats straight home to the refrigerator or freezer.

Storing

Refrigeration

Keep pork below 40 °F during storage. Store uncooked pork items together, separate from cooked foods. Refrigerate or freeze fresh pork Immediately after bringing it home. Never leave meat in a hot car or sitting out at room temperature.

Packaged whole cuts of fresh pork may be refrigerated in their original wrapping in the coldest part of the refrigerator up to four or five days after purchase, while ground pork can be stored in the refrigerator for up to two days. Keep pork refrigerated until you are ready to cook it.

When transporting uncooked or cooked pork to another dining site, place it in an insulated container or ice chest until ready to cook or eat. Cooked pork is at its best when refrigerated no longer than four days.

Table. Safe Handling of Pork in Cold Storage.

Food	Refrigerator (40 °F)	Freezer (0 °F)
Fresh Pork:		
Roast, chops or ribs	3-5 Days	4-6 Months
Ground pork, liver or variety meats	1-2 Days	3-4 Months
Ham (Uncured)	3-5 Days	4-6 Months
Ham (Cured)	5-7 Days	3-4 Months
Cooked Pork:		
Roast, chops, casseroles	3-4 Days	2-3 Months
Ground pork; store-cooked convenience meals	1-2 Days	2-3 Months
Ham (Uncured)	3-4 Days	3-4 Months
Ham (Cured)	3-5 Days	1-2 Months

Freezing

Freeze whole cuts of fresh pork if you do not plan to cook it within four days after purchase. Wrap whole cuts of pork separately in foil or freezer bags before freezing, and label for ease in selecting just the right number of cuts to thaw for a single meal. Be sure to press the air out of the package before freezing.

If you plan to freeze pork in its original wrapping, overwrap the porous store plastic with freezer bag or paper. Cooked pork cuts may be frozen in the same way as fresh, unless made with a sauce or gravy. In that case, pack the meat in a rigid container with a tight-fitting lid.

Preparation

Always wash hands thoroughly with hot, soapy water before preparing foods and after handling raw meat. Don't let raw meat or juices touch ready-to-go foods either in the refrigerator or during preparation. Don't put cooked foods on the same plate that held raw pork. Always wash utensils that have touched raw meat with hot, soapy water before using them for cooked meats. Wash counters, cutting boards and other surfaces raw meats have touched.

Thawing

Thaw uncooked pork in the refrigerator, in cold water or in the microwave oven. Never thaw meat at room temperature. Allow a 24-hour thawing period in the refrigerator. After defrosting raw pork by this method, it will be safe in the refrigerator up to five days before cooking or, if you decide not to use the pork, you can safely refreeze it without cooking it first.

To thaw pork in cold water, leave the meat in its original wrapping or place it in a watertight plastic bag. Change the water every 30 minutes. To thaw pork in the microwave, plan on cooking the meat immediately after thawing because some areas of the food may become warm and begin to cook during microwaving, and any bacteria present wouldn't have been destroyed.

Thawing time will vary according to whether you're thawing a whole roast or cuts and the number of parts frozen together. Use the Defrost or Medium-Low setting, according to the manufacturer's directions. Turn the roast and separate parts as they thaw, taking care the meat does not begin to cook Foods defrosted by the cold water method or in the microwave should be cooked before refreezing because they potentially may have been held at temperatures above 40 °F.

It is safe to cook frozen pork in the oven, or on the stove or grill without defrosting. Estimate one-third to one-half more cooking time depending upon the size of the meat. Broil frozen pork farther away from the heat source; preheat the skillet when pan-frying or pan-broiling. Do not cook frozen pork in a slow cooker.

Marinating

Marinate food in the refrigerator, not on the counter. Discard the marinade after use because it contains raw juices, which may harbor bacteria. If you want to use the marinade as a dip or sauce, reserve a portion before adding raw food, or boil used marinade before brushing on cooked pork.

Partial Cooking or Browning

Never brown or partially cook pork, then refrigerate and finish cooking later, because any bacteria present would not have been destroyed. It is safe to partially pre-cook or microwave pork and lamb immediately before transferring it to the hot grill or oven to finish cooking.

Cooking

One of the critical factors in controlling bacteria in food is controlling temperature. Pathogenic microorganisms grow very slowly at low temperatures, multiply rapidly in mid-range temperatures, and are killed at high temperatures. For safety, foods must be cooked thoroughly. It is essential to use a thermometer when cooking meat and poultry to prevent undercooking and, consequently, prevent food borne illness.

Using a thermometer is the only reliable way to ensure safety and to determine the "doneness" of most foods. To be safe, a product must be cooked to an internal temperature high enough to destroy any harmful bacteria that may have been in the food. Recent research has shown that colour and texture indicators are not reliable. It isn't necessary to rinse raw pork before cooking it. Any bacteria which might be present on the surface would be destroyed by cooking.

Cooking the Meat

For safety, FDA recommends cooking ground pork patties and other ground mixtures to 155 °F. Cook whole muscle meats such as chops and roasts, and fresh cured ham to 145 °F with a 4 minute rest before carving or eating, 160 °F, or 170 °F. Remember that appliances and outdoor grills can vary in heat. Use a meat thermometer to check for safe cooking and doneness of pork. Cooked muscle meats can be pink even when the meat has reached a safe internal temperature. If fresh pork has reached 145 °F throughout, and is given a 4 minute rest, even though it may still be pink in the center, it will be safe. The pink colour can be due to the cooking method or added ingredients. For approximate cooking times of pork.

Internal temperature of safely cooked whole cuts of pork should reach at least 145 °F when measured with thermometer, followed by a 4 minute rest before carving or eating. Roasting Set oven at 350 °F. Roast in a shallow pan, uncovered. Internal temperature: 145 °F (medium-rare) with a 4 minute rest, 160 °F (medium), 170 °F (well-done).

Table. Fresh Pork: Safe Cooking Chart.

Cut	Thickness or Weight	Cooking Time
Loin Roast, bone-in or boneless	2 to 5 pounds	20 to 30 min. per pound
Crown Roast	4 to 6 pounds	20 to 30 min. per pound
Leg, (Fresh Ham) whole, bone-in	12 to 16 pounds	22 to 26 min. per pound
Leg, (Fresh Ham) half, bone in	5 to 8 pounds	35 to 40 min. per pound
Boston Butt	3 to 6 pounds	45 min. per pound
Tenderloin (Roast at 425-450 °F)	½ to 1½ pounds	20 to 30 minutes total
Ribs (Back Country-style or Spareribs)	2 to 4 pounds	1½ to 2 hours (or until fork tender)
Broiling (4 inches from heat) or Grilling		
Loin Chops, bone-in or boneless	¾ inch or 1½ inch	6 to 8 min. or 12 to 16 min.
Tenderloin	½ to 1½ pounds	15 to 25 min.
Ribs (indirect heat), all types	2 to 4 pounds	1½ to 2 hours
Ground Pork Patties (direct heat)	½ inch	8 to 10 minutes—155 °F minimum inte-rnal temperature for all ground meats
In Skillet on Stove		
Loin Chops or Cutlets	¼ inch or ¾ inch 7 to 8 min.	3 to 4 min. or
Tenderloin Medallions	¼ to ½ inch	4 to 8 minutes
Ground Pork Patties minutes	½ inch	8 to 10
Braising: Cover and Simmer with a Liquid.		
Chops, Cutlets, Cubes,	¼ to 1-inch	10 to 25

Medallions		minutes
Boston Butt, Boneless	3 to 6 pounds	2 to 2½ hours
Ribs, all types	2 to 4 pounds	1½ to 2 hours
Stewing: Cover Pan, Simmer, Covered with Liquid.		
Rib, all types	2 to 4 pounds or until tender Cubes	2 to 2½ hours, 1 inch 45 to 60 minutes

Microwaving

When microwaving unequal sizes of pork, arrange in a dish or on a rack so thick parts are towards the outside of dish and thin parts are in the center, and cook on medium-high or medium power. Place a roast in an oven-cooking bag or in a covered pot.

Refer to the manufacturer's directions that accompany the microwave oven for suggested cooking times. Use a microwave-safe thermometer inserted before cooking, or remove meat from microwave and use a digital meat thermometer, to test for doneness in several places to be sure correct temperatures have been reached.

Safe Handling of Poultry

Selecting the Best

Fresh poultry—chicken, turkey, duck and goose—is defined by the USDA as poultry that has never been below 26 °F. When held at temperatures ranging from 26 to 40 °F, there is minimal ice crystal formation and poultry meat is still soft and pliable. There should be no detectable odour.

The surface should not be slick nor have a shiny appearance, and should not have colour defects:

- Look for packages that are cool to the touch and have no tears, punctures, torn or missing labels.
- Always remember to select poultry just before checking out at the register.
- Make sure all poultry, whether raw, prepackaged or from the deli is refrigerated before and after it is purchased.

- Prevent cross-contamination by keeping fresh meats separate from other items. Put raw poultry packages in a plastic bag so "juices" won't drip onto other foods. If raw poultry "juices" do drip on other foods, it is best to throw these other items out and immediately wash surfaces and hands.
- Pack raw poultry in an ice chest if it will take you more than an hour to get home and keep the ice chest in the passenger area of the car during warm weather.
- Take poultry straight home to the refrigerator or freezer.

Dating of Poultry Products

Product dating is not required by federal regulations. However, many stores and processors may voluntarily date packages of poultry or poultry products. If a calendar date is shown, there must be a phrase explaining the meaning of the date. Consumers should use or freeze poultry products within one or two days of purchase.

If the manufacturer has determined a "use-by" date, observe it for peak quality and freshness. It 's always best to buy a product before its date expires. If a date expires after the poultry is frozen, the food can still be used. However, quality begins to deteriorate after one year of frozen storage for whole poultry and nine months of frozen storage for poultry parts.

Product Inspection and Grading

All poultry found in retail stores is either USDA-inspected for wholesomeness and safety, or inspected by state systems using standards equal to the federal government. Prior to entering the processing plant, live birds are inspected for signs of disease. Diseased birds are immediately condemned and do not enter the processing plant.

In the plant, each animal and its internal organs are again inspected for signs of disease. The "Passed and Inspected by USDA" seal ensures that the poultry is wholesome and free from disease. Inspection is mandatory, but grading for quality is voluntary. Poultry are graded according to USDA regulations and standards for meatiness, appearance and freedom from defects. Grade A chickens, the best grade, have plump, meaty bodies and clean skin, free of bruises, broken bones, feathers, cuts and discolouration.

Prestuffed Poultry

Buying retail-stuffed whole poultry is not recommended because of the highly perishable nature of a previously stuffed item. Some USDA-inspected frozen stuffed poultry MUST be cooked from the frozen state to ensure a safely cooked product. Many frozen entrées containing stuffed boneless poultry products may appear to be pre-cooked or browned, but they are NOT ready-to-eat and must be fully cooked by the consumer.

If the label states “Cook and Serve,” “Ready to Cook,” or “Oven Ready” this indicates that the product is raw and must be fully cooked by the consumer. The safest way to cook these products is in a conventional oven. If a microwave oven is used, these poultry products should be covered to allow steam to build, and then allowed to stand for the recommended time to ensure that there are no “cold spots”. Check the internal temperature in several places to make sure the product has been cooked throughout to at least 165 °F

Cooking

Importance of Kitchen Thermometers

One of the critical factors for minimizing bacteria in food is controlling temperature. Most pathogenic microorganisms grow very slowly at low temperatures, multiply rapidly in mid-range temperatures and are killed at high temperatures. For safety, foods must be cooked thoroughly. It is essential to use a thermometer when cooking meat and poultry to prevent undercooking and food borne illness.

Using a thermometer is the only reliable way to ensure safety and to determine the “doneness” of most foods. To be safe, a product must be cooked to an internal temperature high enough to destroy any harmful bacteria that may have been in the food. Colour and texture changes in meat and poultry cannot be relied on to determine that foods have been safely cooked to destroy all bacteria. All poultry and all stuffing, whether cooked alone or in the bird, must be cooked to an internal temperature of at least 165 °F. For reasons of personal taste or texture preferences, consumers may choose to cook poultry to higher temperatures. When cooking whole turkey, the thermometer should be inserted into the thickest part of the thigh, without touching the bone.

For chicken, and other smaller poultry, insert the thermometer in the thickest part of the breast. If stuffed, the center of the stuffing should be checked after the thigh or breast reaches 165 °F. If cooking poultry parts, insert the thermometer into the thickest area, avoiding the bone. The thermometer may be inserted sideways if necessary. When the food being cooked is irregularly shaped, the temperature should be checked in several places. Sometimes consumers will notice a pink colour in fully-cooked poultry. Commercial marinades may have a small amount of nitrites or nitrates which gives poultry a pink colour even when it is fully cooked.

Ground Chicken or Turkey

Ground meats must be cooked thoroughly to kill harmful bacteria. Unlike whole muscle meat, whose interior meat is sterile, the grinding process exposes the interior meat in ground poultry to bacteria, which may be on the surface, in the air, on equipment or on people's hands. To kill bacteria, food safety experts have one major rule of thumb: Cook ground poultry to at least 165 °F. This step, while very simple, offers the best protection that consumers can have for serving ground poultry products safely.

Serving

Wash hands with soap and water before serving or eating food:

- Serve cooked products on clean plates with clean utensils and clean hands. Never put cooked foods on a dish that has held raw poultry unless the dish is washed with soap and hot water.
- Hold hot foods above 140 °F and cold foods below 40 °F.
- Never leave foods, raw or cooked, at room tempe-rature longer than two hours. On a hot day with temperatures at 90 °F or warmer, this decreases to one hour.

Safe Handling of Sausages and Hot Dogs

Selecting the Best Sausage

There are so many varieties of sausages! How long can you store them—and where? Are they fully cooked or not? The following background information will answer these questions and others. Use the storage chart as a guideline for proper handling.

Types of Sausages

Sausages are either ready to eat or not. They can be made from red meat, poultry or a combination. Uncooked sausages include fresh and smoked sausages. Ready-to-eat sausages are dry, semi-dry and/or cooked. Dry sausages may be smoked, unsmoked or cooked. Semi-dry sausages are usually heated in the smokehouse to fully cook the product and partially dry it.

Sausage Labeling Information

Let the label be your guide to sausage selection, handling and—if applicable—cooking. It will list the safe handling and cooking instructions, the nutrient content and the ingredients. Safe handling instructions are mandatory for all raw or partially cooked meat and poultry products. The label must say "Keep Refrigerated" if the sausage is perishable. Product dating is optional but the manufacturer may have affixed a date.

All ingredients in the product must be listed in the ingredient statement in order of predominance from the one weighing the most listed first to the one weighing the least listed last. For sausage products packaged under federal inspection, a Nutrition Facts panel is mandatory.

If sausages are made and packaged in a local store, the nutrient information on the package is voluntary. The Nutrition Facts information on the label can help consumers compare products and make more informed, healthy food choices.

Sausage Defined

Fresh Sausages

Fresh sausages are a coarse or finely ground meat food product prepared from one or more kinds of meat, or meat and meat by-products. They may contain water not exceeding 3 per cent of the total ingredients in the product. They are usually seasoned, frequently cured and may contain binders and extenders.

They must be kept refrigerated and be thoroughly cooked before eating:

- *Fresh Pork Sausages*: May not contain pork by-products and no more than 50 per cent fat by weight.
- *Fresh Beef Sausages*: May not include beef by-products and no more than 30 per cent fat by weight.
- *Breakfast Sausages*: May contain meat and meat by-products and no more than 50 per cent fat by weight.
- *Whole Hog Sausage*: Meat from swine in such proportions as are normal to a single animal and no more than 50 per cent fat by weight.
- *Italian Sausage Products*: Cured or uncured sausages containing at least 85 per cent meat, or a combination of meat and fat, with the total fat content constituting not more than 35 per cent of the finished product. They contain salt, pepper, fennel and/or anise and no more than 3 per cent water. Optional ingredients permitted in Italian sausages are spices and flavourings, red or green peppers, onions, garlic and parsley, sugar, dextrose and corn syrup.

Cooked and/or Smoked Sausages

These products are made of one or more different kinds of chopped or ground meats that have been seasoned, cooked and/or smoked. Water can be no more than 10 per cent by weight. Meat by-products may be used.

Included in this category are:

- Salami
- Bratwurst
- Liverwurst
- Braunschweiger
- Hot Dogs
- Blood Sausage

- Bologna
- Jellied Beef Loaf
- Knockwurst
- Thuringer-Style

Cooked salami is made from fresh meats that are cured, stuffed into casings and cooked in a smokehouse at high temperature. It may be air-dried for a short time. It has a softer texture than dry and semi-dry sausages and must be refrigerated.

Meat Specialties

A ready-to-eat sausage product that is made from finely ground meats that are seasoned and usually cooked or baked rather than smoked. They are usually sliced and served cold.

Included in this category are:

- Chopped Ham Loaf
- Luncheon Meat
- Peppered Loaf
- Head Cheese
- Jellied Corned Beef
- Ham and Cheese Loaf
- Honey Loaf
- Old Fashioned Loaf
- Olive Loaf
- Pickle and Pimento Loaf
- Scrapple
- Souse
- Veal Loaf

Dry and Semi-Dry Sausages

Dry sausages may or may not be characterized by a bacterial fermentation. When fermented, the intentional encouragement of a lactic acid bacteria growth is useful as a meat preservative as well as producing the typical tangy flavour. The ingredients are mixed with spices and curing materials, stuffed into casings, and put through a carefully controlled, long, continuous air-drying process. Dry sausages require more production time than other types of sausage that results in a concentrated form of meat.

Medium-dry sausage is about 70 per cent of its "green" weight when sold. Green weight is the weight of the raw article before addition of added substances or before cooking. Less-dry and fully-dried sausages range from 80 per cent to 60 per cent of original weight at completion.

Dry sausages include:

- Chorizo
- Frizzes
- Pepperoni
- Lola or Lolita and Lyons sausage
- Genoa salami

Semi-dry sausages are usually heated in the smokehouse to fully cook the product and partially dry it. Semi-dry sausages are semi-soft sausages with good keeping qualities due to their lactic acid fermentation.

"Summer Sausage" is the general classification for mildly seasoned, smoked, semi-dry sausages like Mortadella and Lebanon bologna.

Who should Avoid Eating Dry Sausages

Because dry sausages are not cooked, the elderly, very young children, pregnant women and those with weakened immune systems might want to avoid eating them. The bacterium E. coli O157:H7 has been found to survive the process of dry fermenting, and in 1994, some children and adults became ill after eating dry cured salami containing the bacteria.

This is believed to be the first time that this product has been associated with E. coli O157:H7. These illnesses have raised some questions about the effectiveness of processes for producing dry fermented sausage free of this deadly organism. The USDA is looking at ways to identify and correct potential problems in dry sausage products, and is developing procedures for manufacturers to ensure their processing is adequate to destroy bacteria.

Storage of Sausage

All sausage–except dry sausage–is perishable and therefore should be brought directly home when purchased and refrigerated or frozen.

Date on Package of Processed Meats

Although dating is a voluntary programme and not mandated by the federal government, if a date is used it must state what the date means. Since none is a safety date, the product can be used after the date, provided it was stored safely.

Follow the guidelines, for maximum quality in sausage products:

- "Packaging" date is the date of manufacturing, processing, or final packaging."Sell-by" date is the last day a retail store may offer the food for sale. You should buy the product before the date expires, and then use according to the guidelines in the storage chart for maximum quality and safety.
- "Sell-by" date is the last day a retail store may offer the food for sale. You should buy the product before the date expires, and then use according to the guidelines in the storage chart for maximum quality and safety.
- "Best if used by" date tells when the product should be used for best flavour and quality. It is not a purchase or safety date.
- "Use-by" date is the date after which peak quality of the product begins to decrease, but the product may still be used.
- "Expiration" date marks the end of the product's useful life or the last day to be used.

Selecting the Best Hot Dogs

Types of Hot Dogs

Whether you call it a frankfurter, hot dog, wiener or bologna, it's a cooked sausage and a summertime favourite. They can be made from beef, pork, turkey or chicken–the label must specify which.

All ingredients in the product must be listed in the ingredient statement in order of predominance from the one weighing the most listed first to the one weighing the least listed last. And there are federal standards for their content. Smoking and curing ingredients contribute to flavour, colour and preservation of the product. They come in all shapes and sizes–short, long, thin and chubby.

The most popular of all categories, the skinless varieties, have been stripped of their casings after cooking. Water or ice may be used to facilitate chopping or mixing or to dissolve curing ingredients. Sausages may contain no more than 10 per cent water and 30 per cent fat or a combination of 40 per cent fat and added water. Up to 3.5 per cent nonmeat binders and extenders such as nonfat dry milk, cereal, dried whole milk or 2 per cent isolated soy protein may be used, but must be shown in the ingredient statement by its common name.

By-Products, Variety Meats

Frankfurters, hot dogs, wieners or bologna "with by-products" or "with variety meats" are made according to the specifications for cooked smoked sausages except they consist of not less than 15 per cent of one or more kinds of raw skeletal muscle meat with raw meat by-products. The by-products must be accompanied by the name of the species from which it was derived and must be individually named in the ingredient statement.

Species

Beef franks or pork franks are cooked, smoked sausage products made according to the specifications above, but with meat from a single species and do not include by-products. Turkey franks or chicken franks contain turkey or chicken skin and fat in natural proportions of that found on a turkey or chicken carcass.

Mechanically Separated Meat or Poultry

Carcass parts from which most of the meat has been removed still have usable meat attached. These parts are pushed under high pressure through equipment with openings so fine that a small amount of powdered bone the size of a grain of sand may pass through along with the remaining muscle meat and other soft tissue. This is called "mechanically separated" meat, and if used in a product, the label must state it. If a serving contains 20 mg or more of calcium from the finely powdered bone, the label must give the calcium content as a percentage of the US RDAs.

Mechanically Deboned Poultry

This does not have the same requirements as mechanically separated meat and is simply listed in the ingredients statement as "chicken" or "turkey."

Handling of Hot Dogs

When you leave the grocery store with any kind of sausage, head straight home and refrigerate or freeze it immediately. If there is a date on the package, follow those guidelines for use. If there is no date, hot dogs can be safely stored unopened in the refrigerator for up to two weeks. Once opened, they are safe in the refrigerator for only one week. For maximum quality, freeze hot dogs no longer than one to two months. Never leave hot dogs at room temperature for more than two hours, or in the hot summer months when the temperature rises to 90 °F or above, for more than one hour. Finally, even though hot dogs are fully cooked, if you choose to reheat them

Freeze the product if you cannot use it within the times recommended above for refrigerator storage. Once frozen, it does not matter if the date expires, because all foods kept frozen continuously are safe indefinitely. However, for best quality, use within 1 to 2 months.

Safe Handling of Veal

What is Veal

Veal is the meat from a calf or young beef animal. A veal calf is raised until about 16 to 18 weeks of age, weighing up to 450 pounds. Male dairy calves are used in the veal industry. Dairy cows must give birth to continue producing milk, but male dairy calves are of little or no value to the dairy farmer. A small percentage are raised to maturity and used for breeding. A calf is a young bovine of either sex that has not reached puberty and has a maximum live weight of 750 pounds. "Bob" veal is a veal calf marketed up to 3 weeks of age or at a weight of 150 pounds. "Special-Fed" veal is usually fed nutritionally balanced milk or soy based diets. These specially controlled diets contain iron and 40 other essential nutrients. The majority of veal calves are "special-fed".

Selecting the Best

Choose veal in the fresh meat case that is grayish pink in colour and firm to the touch. Vacuum packaged veal in the self-serve case will be more maroon in colour because it has not been exposed to oxygen. Look for packages that are cool to the touch, have no wear or punctures, and little or no excess liquid. Always check the "sell-by" date. Purchase raw meats last.

Make sure all meats-whether raw, pre-packaged or from the deli are kept refrigerated. Fresh meats may contaminate other grocery items. The best way to prevent this cross-contamination is to always keep fresh meats from other items. Put raw meat packages in a plastic bag so juices won't drip onto other foods. Pack raw meats in an ice chest if it will take more than an hour to get home. Keep ice chest in the passenger area of the car during warm weather. Take meats straight home to the refrigerator or freezer.

Product Dating

Product dating, applying "sell-by" or "use-by" dates, is not required by federal regulations. However, many stores and processors may voluntarily choose to date packages of raw veal or processed veal products. Use or freeze products with a "sell-by" date within three to five days of purchase. If the manufacturer has determined a "use-by" date, observe it. It's always best to buy a product before its date expires. It's not important if a date expires after freezing veal, because all foods stay safe while properly frozen.

Product Inspection and Grading

All veal found in retail stores is either USDA-inspected for wholesomeness or inspected by state systems which have standards equal to the federal government. Each animal and its internal organs are inspected for signs of disease. The "Passed and Inspected by USDA" seal ensures that the veal is wholesome and free from disease. Although inspection is mandatory, its grading for quality is voluntary, and a plant pays to have its veal graded. Quality grade refers to the eating quality of the meat. Veal and calf carcasses are graded on a composite evaluation of two general grade factors: conformation and quality of the lean. In addition, the colour of the lean carcasses is key in differentiating between veal, calf and beef carcasses. The five grades for veal are as follows: prime, choice, good, standard and utility.

Retail Cuts

There are seven basic major cuts into which veal is separated: leg, sirloin, loin, rib, shoulder, foreshank and breast. When examining a package of veal, the label can help the purchaser identify the meat in the package. For example, a label stating "veal rib chop" identifies the packaged meat as "veal", the primal or large wholesale cut from the "rib", and the name of the retail cut as "chop".

This information helps consumers know what type of preparation method to use. The most readily available cuts of veal today include rib chops, loin chops, veal for stew, cutlets, arm steak, blade steak, rib roast, breast, shanks and round steak.

Storage

Refrigeration

Keep veal below 40 °F during storage. Store uncooked veal items together, separate from cooked foods. Refrigerate or freeze fresh veal Immediately after bringing it home. Never leave veal in a hot car or sitting out at room temperature. Packaged whole cuts of fresh veal may be refrigerated in their original wrappings in the coldest part of the refrigerator for three to five days after purchase, while ground veal can be stored in the refrigerator for one or two days.

Keep veal refrigerated until you are ready to cook it. When transporting cooked veal to another dining site, place it in an insulated container or ice chest until ready to eat. Cooked whole cuts of veal are at their best when refrigerated no longer than two to three days. Cooked ground veal is best when refrigerated no longer than one or two days.

Freezing

Freeze whole cuts of fresh veal if you do not plan to cook it within three to five days after purchase. Freeze ground veal if you do not plan to cook it within one to two days after purchase. Wrap veal parts separately in aluminum foil or freezer paper before freezing, and label for ease in selecting just the right number of parts to thaw for a single meal. Be sure to press the air out of the package before freezing. If you plan to freeze veal in its original wrapping, overwrap with freezer bag or paper. Cooked parts may be frozen in the same way as fresh, unless made with a sauce or gravy; in that case, pack in a rigid container with a tightfitting lid.

Preparation

Cleanliness

Always wash hands thoroughly with hot soapy water before preparing foods and after handling raw veal. Don't let raw meat or juices touch ready-to-go foods either in the refrigerator or during preparation. Don't put cooked foods on the same plate that held raw veal. Always wash utensils that have touched raw meat with hot, soapy water before using them for cooked meats. Wash counters, cutting boards and other surfaces raw meats have touched.

Thawing

Thaw uncooked veal in the refrigerator or in cold water. Never thaw veal at room temperature. Allow a 24-hour thawing period. After defrosting raw veal by this method, it will be safe in the refrigerator for up to five days before cooking, or, if you decide not to use the veal, you can safely refreeze it without cooking it first. To thaw veal in cold water, leave the veal in its original wrapping or place it in a watertight plastic bag. Change the water every 30 minutes.

For quick thawing of uncooked or cooked veal, use the microwave, but plan on cooking the meat immediately after thawing, because some areas of the food may become warm and begin to cook during microwaving. Thawing time will vary according to whether you're thawing a whole roast or cuts and the number of parts frozen together. Use the Defrost or

Medium-Low setting, according to the manufacturer's directions. Turn the roast and separate parts as they thaw, taking care the meat does not begin to cook. Repeat as needed. Foods defrosted by the cold water method or in the microwave should be cooked before refreezing, because they may have been held at temperatures above 40 °F.

Table. Safe Handling of Veal in Cold Storage.

Product	Refrigerator (40°F)	Freezer (0°F)
Raw Whole Cuts of Veal: Chops, roasts and steaks	3-5 Days	4-6 Months
Raw Ground Veal and Stew	1-2 Days	3-4 Months
Broiled, Fried, Grilled or Roasted Veal	2-3 Days	2-3 Months
Cooked Ground Veal and Gravies Made From Veal	1-2 Days	2-3 Months

It is safe to cook frozen veal in the oven, or on the stove or grill without defrosting. Estimate one-third to one-half more cooking time depending upon the size of the meat. Broil frozen veal farther away from the heat source; preheat the skillet when pan-frying or pan-broiling. Do not cook frozen veal in a slow cooker.

Marinating

Marinate food in the refrigerator, not on the counter. Boil used marinade before brushing on cooked veal. Discard any uncooked, leftover marinade after use because it contains raw juices, which may harbor bacteria.

Rinsing

There is no need to rinse raw veal before cooking. Any bacteria that might be present on the surface would be destroyed by cooking.

Partial Cooking or Browning

Never brown or partially cook veal, then refrigerate and finish cooking later, because any bacteria present would not have been destroyed. It is safe to partially precook or microwave veal IMMEDIATELY before transferring it to a hot grill or oven to finish cooking.

Cooking

Importance of Kitchen Thermometers

One of the critical factors in controlling bacteria in food is controlling temperature. Pathogenic microorganisms grow very slowly at low temperatures multiply rapidly in mid-range temperatures, and are killed at high temperatures. For safety, foods must be cooked thoroughly. It is essential to use a meat thermometer when cooking veal to prevent undercooking and, consequently, prevent food borne illness. Using a thermometer is the only reliable way to ensure safety and to determine the "doneness" of most foods.

To be safe, a product must be cooked to an internal temperature high enough to destroy any harmful bacteria that may have been in the food. Recent research has shown that colour and texture indicators are not reliable. When cooking whole cuts or parts of veal, the thermometer should be inserted into the thickest part of the meat, away from the bone, fat and gristle. The thermometer may be inserted sideways if necessary. When the food being cooked is irregularly shaped, the temperature should be checked in several places.

Whole Muscle Meats

Cook whole cuts of meat to a minimum internal temperature of 145 °F for medium-rare, with a 4 minute rest before carving or eating or 160 °F for medium-cooked whole cuts of meat, and 170 °F for well-done cuts.

Ground Veal

Ground veal must be cooked thoroughly to kill harmful bacteria. Unlike whole muscle meat that is sterile inside, the grinding process exposes the interior meat in ground veal to bacteria that may be on the surface, in the air, on equipment or on people's hands. To kill these bacteria, food safety experts have one major rule of thumb—cook ground veal to at least 155 °F. This step, while very simple, offers the best protection that consumers have to serve ground veal safely.

Microwaving

When microwaving unequal size pieces of veal, arrange in a dish or on a rack so thick parts are towards the outside of the dish and thin parts are in the center, and cook on medium-high or medium power. Place a roast in an oven-cooking bag or in a covered pot. Refer to the manufacturer's directions that accompany the microwave oven for suggested cooking times. Use a microwave-safe thermometer inserted before cooking, or removes meat from microwave oven and use a digital meat thermometer in several places to be sure correct temperatures have been reached. Meats cooked in the microwave oven must be cooked to a minimum internal temperature of 165 °F and allowed to stand covered at least 2 minutes. Follow standing or rest times given.

References

Bhupendra Singh Khatkar.: *Food Science and Technology*, Daya Publications, Delhi, 2007.

Bimal Kumar Kapoor.: *Food Science and Nutrition*, Mohit Publications, Delhi, 2007.

Dev Raj.: *Food Science and Technology: Glossary of Preeminence*, New India Publications, Delhi, 2011.

4

Food Storage

Food caches, Hooper Bay, Alaska, 1929. Photograph by Edward S. Curtis Food storage is both a traditional domestic skill and is important industrially. Food is stored by almost every human society and by many animals.

Storing of food has several main purposes: preparation for periods of scarcity or famine taking advantage of short term surplus of food as at harvest time enabling a better balanced diet throughout the year preparing for special events and celebrations planning for catastrophe or emergency protection against predators or others.

Domestic Food Storage

Grain

Grain is stored in rigid sealed containers to prevent ingress of moisture or attack by vermin. For domestic quantities metal cans are used (in the USA the smallest practical grain storage uses closed-top #10 metal cans).

Storage in grain sacks is not effective. Mold and pests destroy a 25kg cloth sack of grain in a year — even if stored off the ground in a dry area. On the ground or damp concrete, the time is as little as three days, and the grain might have to be dried before it can be milled. Food stored under unsuitable conditions should not be purchased or used because of risk of spoilage. To test whether grain is still good, sprout some. If it sprouts, it is still good, but if not, it should not be eaten. It may take up to a week for grains to sprout. When in doubt, throw it out.

Meat

Unpreserved meat has only a relatively short life in storage. Pork should be eaten within one day but beef and venison improve with up to 5 days storage in a cold room. Dry aging techniques are sometimes used to tenderize specialty gourmet meats by hanging them in carefully controlled environments for up to 21 days. Semi-dried meats like salamis and country style hams are processed first with salt, smoke, sugar, or acid, or other "cures" then hung in cool dry storage for extended periods, sometimes exceeding a year.

Fish and Shellfish

It is unsafe to store fish or shellfish without preservation. Fresh shellfish and whitefish should be eaten within a few hours of harvesting.

Use of Stored Food

Guidance for surviving emergency conditions in many parts of the world recommends acquiring a limited range of grains (usually corn, wheat and beans supplemented with oil, dried milk, and vitamins) and then preparing them in simple ways for long-term survival. This may not be wholly practical because of appetite exhaustion.

An unvarying diet of staples prepared in the same way causes most people to eat less. Garden-grown fruits and vegetables, freeze-dried, canned, and fresh-baked foods are essential supplements to such a program. A special virtue of home stored foods is their low cost. Costs of dry bulk foods (before preparation) are often less than 1/4 of convenience and fresh foods purchased at supermarkets.

Commercial Food Storage

Grain and beans are stored in tall grain elevators, almost

always at a rail head near the point of production. The grain is shipped to a final user in hopper cars. In the former Soviet Union, where harvest was poorly controlled, grain was often irradiated at the point of production to suppress mold and insects. In the U.S., threshing and drying is performed in the field, and transport is nearly sterile and in large containers that effectively suppress pest access, so irradiation is not required. At any given time, the U.S. usually has about two weeks of stored grains.

Fresh fruits and vegetables are either packed in plastic cups in cardboard boxes for fresh premium markets, or placed in large plastic tubs for sauce and soup processors. Fruits and vegetables are usually refrigerated at the earliest possible moment, and even so have a shelf life of two weeks or less.

There is a thriving but small market in bulk vegetables and convenience foods for campers.

In the USA meat animals are usually transported live, slaughtered at a major distribution point, hung and transported for two days to a week in refrigerated rail cars, and then butchered and sold locally. Before refrigerated rail cars, meat had to be transported live, and this placed its cost so high that only farmers and the wealthy could afford it every day. In Europe much meat is transported live and slaughtered close to the point of sale. In much of Africa and Asia most meat is for local populations is reared , slaughtered and eaten locally which is believed to be much less stressful for the animals involved and requires very little meat storage capacity. In Australia and New Zealand where a large proportion of meat production is for export meat is stored in very large freezer plants before being shipped overseas in freezer ships.

References

Dilip Shah: *Co-Operativization Liberalization and Dairy Industry in India*, ABD Publishers, Delhi, 2000.

Doreen Virtue: *Constant Craving: What Your Food Cravings Mean And How To Overcome Them*, Hay House India, 2011.

Ernest R. Vieira.: *Elementary Food Science*, Chapman and Hall, 2010.

5

Packaging of Fresh and Processed Meat

Introduction

Meat packing is the wholesale slaughtering, processing and packaging of animals into meats to distribute to retailers such as supermarkets. The wholesale meat packaging industry purchases the animals from farms. Meat is considered to be the flesh of pigs, cows, lambs and sheep. The flesh of chickens, ducks and other poultry is not considered to be meat; fish is not considered to be meat either. Meat packing is a worldwide industry that may or may not be inspected and controlled by government agencies, depending on each location.

Safety and sanitation should be important concerns in meat packing since disease and contamination of meats can occur in unsanitary conditions. Since meat is the flesh of slaughtered animals, it must be kept cold as soon as possible after the slaughter so it doesn't spoil. Proper refrigeration and freezing are crucial in the meat industry.

The earliest recorded commercial distribution of frozen meat was in 1874. The meat was shipped to England from the United States. Chicago remained America's largest meat packing region until the 1920s. In the meat industry's early days, packing plants had rooms cooled with ice to store meats. Cattle shipper Gustavus Swift is credited with the development of railroad car cooling systems using ice.

This made shipping meat over long distances by railway possible before electric refrigeration systems were invented. Meat packing plants may specialise in a certain type of meat or product. For instance, some commercial meat packers only pack, process and distribute organic meats. Organic meat comes from animals that aren't given any antibiotics or growth hormones. Governing bodies are formed to routinely inspect farms that raise animals for organic meat. Other packing companies only export frozen meats or just process one type, such as sheep meat from New Zealand.

The meat packing industry tends to have many critics. Issues of sanitation as well as animal cruelty and worker abuse surface from time to time. Diseases such as the mad cow epidemic in the beef industry do occasionally occur in meat plants. Meat industry critics have complained of low pay for immigrant workers as well as butchers getting burned, cut and loosing fingers.

Most countries have at least some government licensing, employment and inspection standards, but these may be inadequate in some areas. Those opposed to meat packing practices in some parts of the world have protested to try and change conditions. Some governments have responded with closer monitoring and inspection procedures. The function of packaging is to surround or wrap meat products with suitable protective material. Packaging materials were in the old days simple natural materials, *e.g.*, leaves, but nowadays exclusively manufactured materials such as paper or synthetic films.

Purpose of packaging

The basic purpose of packaging is to protect meat and meat products from undesirable impacts on quality including microbiological and physio-chemical alterations. Packaging protectsfoodstuffs during processing, storage and distribution from:

- Contamination by dirt (by contact with surfaces and hands)
- Contamination by micro-organisms (bacteria, moulds, yeasts)
- Contamination by parasites (mainly insects)
- Contamination by toxic substances (chemicals)
- Influences affecting colour, smell and taste (off-odour, light, oxygen)
- Loss or uptake of moisture (evaporation or water absorption)

Adequate packaging can prevent the above listed secondary contamination of meat and meat products. But the further growth of microorganisms, which are already present in meat and meat products, cannot be interrupted through packaging only. To halt or reduce microbial growth, packaging has to be combined with other treatments, such as refrigeration, which will slow down or stop the further growth of microorganisms, or with heating/sterilisation, which will reduce or completely eliminate contaminating microorganisms.

Fig. Meat Products in Different Packages of Synthetic Materials.

The packaging procedure results in an inner package, where the packaging material is in direct contact with the product. In some cases it is combined with an outer package often a cardboard boxe, or other materials.

There are various synthetic packaging films available for the inner packaging, *e.g.*, transparent or opaque, flexible or semi-rigid, gas proof or permeable to certain gases. These materials are selected to serve specific purposes, such as protection from unwanted impacts or attractive presentation.

Requirements for Packaging Materials

A range of synthetic materials suitable for meat packaging are available mainly in the form of plastic films or foils.

Packaging films must be/have:

- Flexible
- Mechanical strength
- Light weight

- Odourless
- Hygienic (clean and toxicologically harmless)
- Easy recycling
- Resistance to hot and cold temperatures
- Resistance to oil and fats
- Good barrier properties against gases
- Sealing capability
- Low-cost

Barrier Against Gases

Good barrier properties against oxygen and evaporation are the most important features in order to ensure:

- *Exclusion of oxygen:* Air contains about 20 percent oxygen. Oxygen negatively affects unpackaged meat and meat products during prolonged storage periods. It changes the red meat colour to grey or green and causes oxidation and rancidity of fats resulting in an undesirable off-flavour.

  The oxygen permeability of films used for the packaging of meat products varies. The lower the oxygen permeability the more efficient the protection of product quality. The best protection will be achieved using oxygen-proof packaging films together with vacuum packaging of the product. This ensures that practically no oxygen remains in the package and no oxygen will penetrate from the air into the packaged product.

  While oxygen is generally undesirable in packages of meat and meat products, there is one exception where oxygen-permeable foils are desirable, namely for fresh ready-to-sell meat portions in self-service outlets. In this specific case the utilisation of oxygen-permeable foils produces a desirable bright red meat colour.

- *Prevention of evaporation of product moisture:* Fresh meat or fresh sausages, cooked ham, etc., have a relatively high moisture content and will suffer considerable weight and quality losses by evaporation and drying during storage if such products remain unpacked. The packaging material must therefore be sufficiently water-vapour-proof.

Barrier Against Light

The prolonged exposure of meat and meat products to daylight or artificial light accelerates unattractive colour changes, oxidation and rancidity because light provides the energy for these processes. Transparent packaging films normally used for meat products allow attractive product presentation as the packaged product is visible.

However, such films provide no protection against light impact. Normally products in transparent packaging films are sufficiently protected when stored in the dark or moderate llight conditions. For light sensitive products or products exposed to strong light, coloured or opaque films should be used. Films laminated with aluminium foil are very effective.

Fig. Opaque and Printed Films as Light Barrier.

Fig: Aluminium Foil Bag.

Sealing Capability

Some packaging materials are required to have good thermoplastic properties. They are heat sealable, which means that two of these films, put closely in contact to each other under slight pressure and with simultaneous high temperature application, will melt or seal together along the heated area, resulting in hermetically closed plastic pouches or bags.

Types of packaging films

Practically all films used for meat packaging derive from synthetic "plastic" materials.

Cellulose, which is not a synthetic but a natural material derived from wood, was formerly widely used in the form of transparent films.

It is now no longer of great importance in meat packaging although still used for some specific purposes. However, cellulose is still important for the manufacture of certain kinds of artificial sausage casings.

The most common synthetic materials used for meat packaging are:

Polyethylene (PE)	(oxygen +, water vapour -)
Polypropylene (PP)	(oxygen +, water vapour -)
Polyvinylchloride (PVC) (soft)	(oxygen +, water vapour -)
Polyester (PET)	(oxygen ±, water vapour -)
Polyamide (PA)	(oxygen -, water vapour +)

+ = relatively permeable

- = relatively impermeable

$$\left.\begin{array}{l}\text{Polyvinylidenchloride (PVDC)}\\ \text{Ethylenvinyl alcohol (EVOH)}\end{array}\right\}\text{used as barrier plastics}$$

Foils made from the above synthetic materials are selected based on their different properties related to oxygen and water vapour.

For the various purposes in the meat industry packaging films can be divided into

- Single-layer films or
- Multi-layer films

Fig. Single-layer Film.

Single-Layer Films

One common use of single-layer films is the wrapping of meat pieces, processed meat products, bone-in or boneless meat cuts or even entire carcasses. These films are usually self-adhesive, *i.e.*, they cling together -"cling film"- in the overlapping areas. Hence they provide good protection from external contamination and to some extend from evaporation, but no protection from oxygen, as they are not hermetically closed or sealed packages. Foils with good self-adhesive properties are PE, PA, PVC and PP.

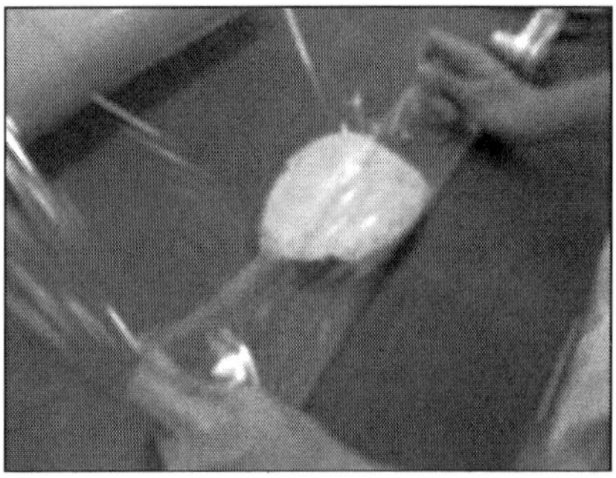

Fig. Wrapping with Single-layer Film.

Another important utilisation for single-layer films is in freezer storage. For meat blocks, meat cuts or smaller portions of meat or meat products, single-layer films are stretched tightly around the meat surface before freezing. The tight film prevents evaporation losses, which occur during freezer storage of unpacked products. The film is in tight contact with the products surface, in order to avoid evaporation, ice formation and freezer burn at non- contact spots. Suitable cold resistant films for freezer storage are PA or PE.

Fig. Freezer Burn Developed During Freezer Storage because of Insufficient Protection against Evaporation.

One specific utilisation for single-layer films is the wrapping of chilled meat portions for self-service outlets (supermarkets, etc.). Those meat pieces (beef, pork or chicken) are placed in a hygienic cellulose or plastic tray and tightly wrapped with single-layer plastic film. The ends of the foil are overlapped at the bottom side of the tray, where they firmly cling together. Films to be used should have low water vapour permeability to avoid the drying out of the meat during storage.

But for making attractive to customers, such meat needs to retain an attractive bright red meat surface colour (oximyoglobin) especially in the case of fresh portioned beef. For this reason the plastic foils to be used shall have a high oxygen permeability so that the oxygen of the air can react with the myoglobin of the meat and form the bright red oximyoglobin.

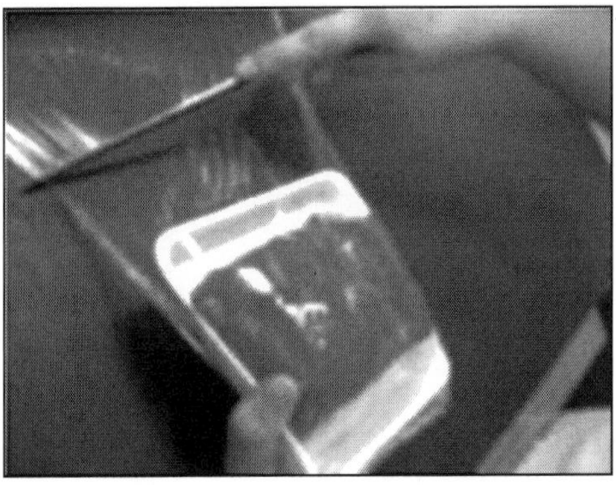

Fig. Fresh Meat being Packed in Self-service Tray.

Oximyoglobin is not a chemical compound but a loose aggregation of oxygen to the red meat pigment myoglobin, which keeps meat bright red for a number of hours. Suitable single-layer films for fresh meat packaging are PE or soft PVC. Formerly cellulose films were also used, which have the same permeability pattern but are less self-adhesive and the overlapping ends do not cling together very well.

Multi-Layer Films

Practically all the other films used for meat packaging are designed as strong oxygen and water-vapour barriers. In order to fully achieve these requirements, films with good barrier properties for oxygen and water vapour respectively are combined.

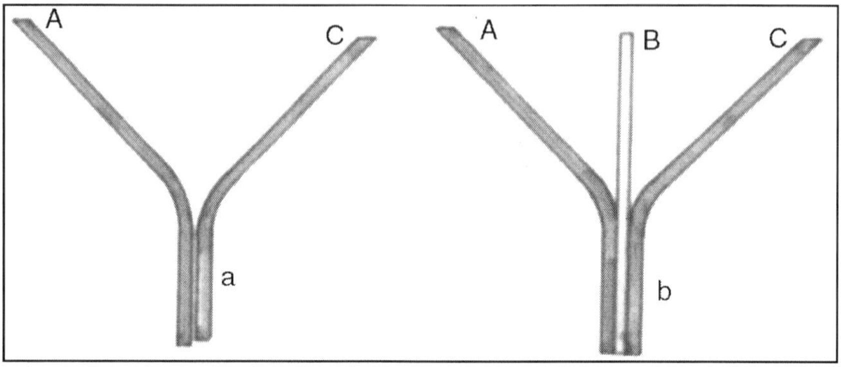

Fig. Typical Multi-layer Films (a = two layers, b = three layers).

- Layer A: Outside layer (mechanically strong, gas barrier to oxygen)
- Layer B: Middle layer (barrier to oxygen)
- Layer C: Inside layer = sealant layer (capable of being melted and welded under pressure to the sealant layer of the opposite sheet of the bag/pouch, serves also as barrier to water vapour)

A very efficient combination is PA/PE. PA is used as the outside layer for example for films for vacuum bags. PA is relatively oxygen proof but permeable to some extend to water vapour.PE has the opposite properties, it is water vapour proof but permeable to oxygen. The combination of both renders such a multi-layer film very tight against oxygen and water vapour evaporation. Moreover, the PE used as the inside layer has good thermoplastic properties and is therefore well suited for heat sealing. The PA/PE combination is the most simple structure for a multi-layer film. The packaging industry has refined the systems by introducing additional layers which serve as strong oxygen barriers. Sealant layers consist mostly of Polyethylene (PE) or Ionomer (I). Outside layers may be Polyamide (PA), Polyester (PET) or Polypropylene (PP). Barrier layers for oxygen are made of Polyvinylidenchloride (PVDC) or materials with similar properties.

Fig. Sausage Slices Placed in Vacuum Bag Prior to Sealing.

Vacuum bags, used for vacuum packaging machines are composed of two or more sheets of multi-layer films. By drawing the vacuum and sealing of such bags, the air is excluded from the package and the damaging effects of oxygen such as rancidity or discolouration of the packed products will be significantly slowed down or not develop at all.

However, exposure to strong light may cause discolouration even under vacuum.

Fig. Start of Evacuation in Vacuum Packaging Machine.

Fig. Evacuated and Sealed Product.

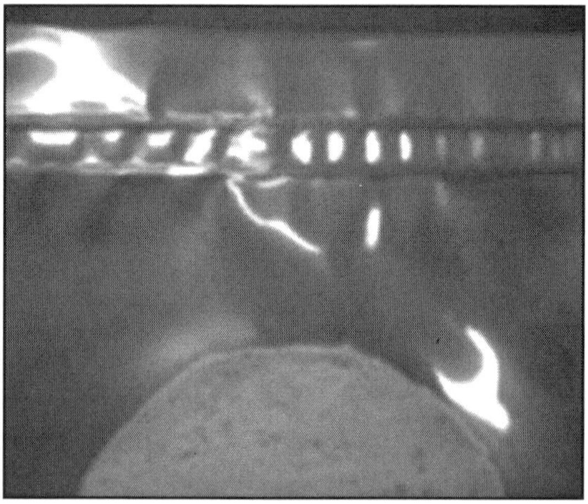

Fig. Good Sealing Seam.

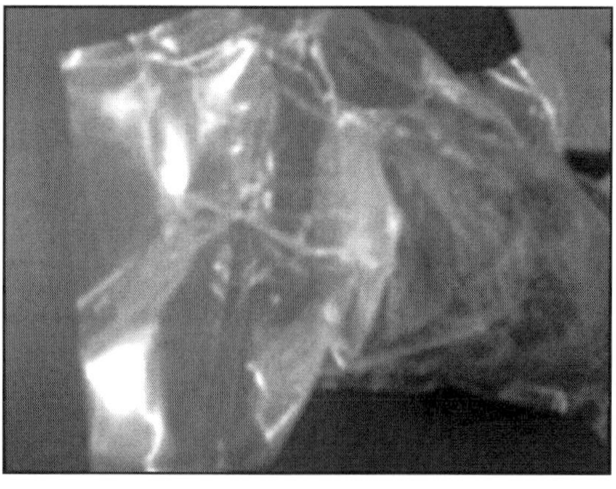

Fig. Faulty Sealing Seam: Risk of Air Penetrating Inside the Package because of Film Overlapping.

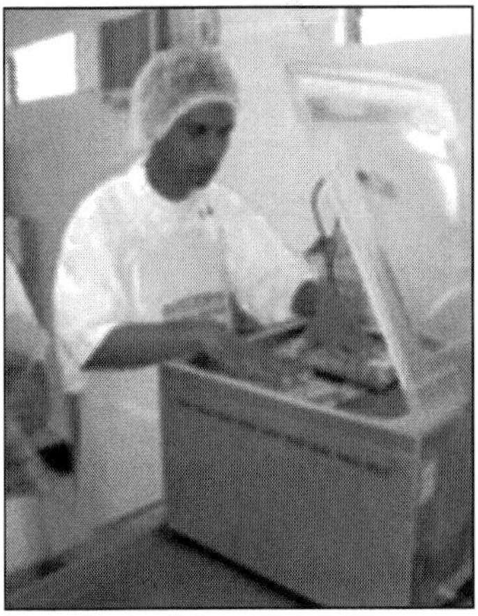

Fig. Simple Vacuum.

Processed meat products in slices or as entire pieces are packed in small to medium-size vacuum bags. For larger sized products, bags made of shrinkable films can be used where, after vacuum-packaging, the product in its package of synthetic film is sprayed with or dipped into hot water (80°C). The contact with the hot water causes the shrinkage of the thermoplastic film and results in tight impermeable wrapping of the goods. Shrink films may for example be composed as follows: PET/PA/EVOH/PO

Fig. Vacuum Packed Beef Cuts/ Rump, Tenderloin, Roast Beef.

Vacuum packaging for fresh chilled boneless beef cuts is mainly done to promote meat ripening. The beef meat cuts can remain for a number of weeks (maximum 3 months) in the vacuum bag, provided oxygen remains excluded and the storage temperature is kept at -1°C (which is just above the freezing point of meat). During these storage conditions lactic acid bacteria (which do not cause spoilage!) will inhibit the growth of most other bacteria, which results in the prolonged microbial stability. Beef becomes very tender during such extended ripening periods without losing much of its typical flavour.

Heat treatment or cooking for some meat products can be carried out in the package after vacuum-packing. Temperatures of 60°-80°C or even higher up to sterilisation temperatures (above +100°C) can be employed for hams, sausages, etc. In these cases a pasteurisation or sterilisation effect of the uncooked packaged products is achieved and re-contamination avoided as long as the package is not opened.

For specific products such as entire sausages, semi-automatic vacuum-packaging can be employed. A bottom film is moulded according to the shape of the sausages by using heat and force (by compressed air or mechanical). These machines are called thermo-formers. The sausages are loaded and a rigid top film is sealed on after evacuating the moulded spaces. Individual product portions are cut apart along their sealing layers.

Fig. Semi-automatic Vacuum Packaging. Loading Sausages in Moulded Bottom Film. Top Film to be Sealed on Bottom Film after Drawing Vacuum. Individual Portions to be Cut apart.

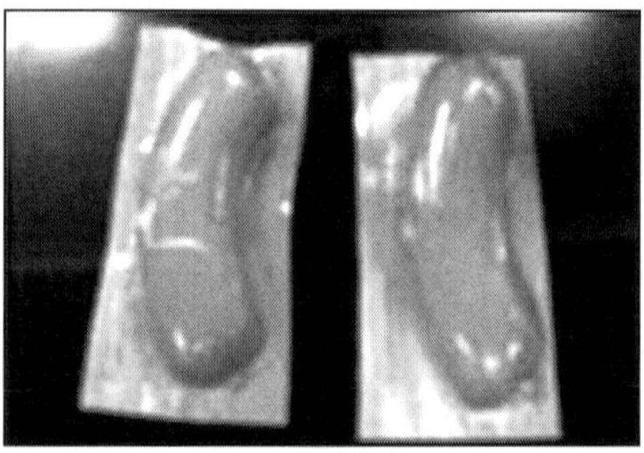

Fig. Product Vacuum Packed in Moulded Bottom Film and Rigid Top Film.

A packaging method commonly used in larger meat industries is skin packaging. For this method the products are placed in the packaging machine, usually on a rigid film, which serves as the bottom layer of the final package. Another flexible film (top layer, which is heated for increased flexibility) drapes itself from above around the product, resembling a tight "skin" on the product surface avoiding wrinkles and purges. The skin-like coverage of the product takes place in a sealing station in the packaging machine, where the top and bottom film are sealed around the edges. Individual packages are separated by cutting around the bottom seal perimeter.

Fig. Skin Packaging.

The latest development in this sector is the "form-shrink" packaging technology. Products, *e.g.*, meat cuts, chicken carcasses, entire sausages, smaller portions of meat products, are placed between two shrinkable films, which are moulded without wrinkles around the goods. Sealing seams can be kept extremely small. This technology is very cost-effective in terms of usage of packaging films but requires high-tech equipment and is only of relevance for large-scale industries.

A useful technology is Modified Atmosphere Packaging (MAP) of meat and meat products. The packaging materials used are gas-proof multi-layer films composed for example of PE, PA and barrier layers. Rigid films may be used to mould cup or box shaped containers which are filled. A flexible lid foil is then sealed on. MAP packaging can also be done for ordinary plastic bags/pouches.

Fig. Skin Packaging.

MAP packages are firstly subjected to a vacuum. A mixtures of gases is the introduced into the air-free space before sealing. The gas mixture usually contains nitrogen (N_2) and carbon dioxide (CO_2). N_2, which is also the major constituent of atmospheric air, is inert, *i.e.*, it does not react with meat product components such as fat or myoglobin. Its function is to replace the atmospheric oxygen (O_2) and thus prevents O_2 induced negative impacts. The other component of the gas mix, CO_2, has a protective function, as it inhibits to some extend the growth of bacteria and moulds.

Fig. Modified Atmosphere Packaging (MAP). Lid Foil of One Package Segment Opened (right).

The gas mixture commonly used is 20 per cent-30 per cent CO_2 and 70 per cent-80 per cent N_2. This is applicable for all processed meat products. If fresh meat pieces are to be packed in gas-proof packages instead of wrapping them with oxygen-permeable foil, the bright-red fresh meat colour can be achieved by adding oxygen to the gas mix to be injected into the package and replacing the N2 content accordingly. As sufficient oxygen is needed to maintain the bright-red colour, gas mixes for fresh meat are usually composed of 70 per cent-80 per cent O_2 and 20- per cent-30 per cent CO_2.

Skin- and MAP-packaging are often too sophisticated for the small producers, but may be of increasing interest to medium-size meat plants. There are now small manual and semi-automatic packaging machines available, which are designed for smaller throughputs. However, the utilisation of those machines implies that the necessary types of synthetic films, and in the case of MAP-packaging, also the relevant gases, are available.

Meat-Packing Industry

Meat-Packing Industry, large industry involving the slaughtering, processing, and distribution of cattle, sheep, and hogs. It is one of the most important industries in the U.S., and has its primary concentration in the Middle West. The packing industry has tended to decentralise in recent years, and slaughtered livestock are now generally moved directly from farms, ranches, and feedlots to meat-packers. The cattle-slaughtering sector of the industry, in particular, has become concentrated in cattle-raising regions to the west-namely the western Corn Belt and the Great Plains, where beef is shipped to wholesalers and retailers primarily in the form of fresh primal cuts. Hog slaughtering is still carried on chiefly in large plants, where the hogs are processed into numerous cuts and products.

In accordance with the Humane Methods of Slaughter Act of 1978, all livestock are now made insensible before they are killed. For cattle and sheep, a captive bolt, a type of gun designed for stunning, is generally used. Hogs often are immobilised painlessly through gassing.

Many parts of the slaughtered animals are shipped for consumption as fresh meat; other parts, especially of the hog, are cured and smoked. The fatty portions are converted into lard and commercial grease by rendering processes. Bones are converted into glue, fertilizer, animal feeds, and other usable products, including pharmaceuticals; hoofs and horns are used or sold for other purposes.

Refrigeration and Transportation

The meat of slaughtered livestock, after thorough chilling, is shipped in refrigerated motor carriers to cities. There the meat is delivered to wholesale distribution centers and grocery centers. Before cold storage of meat was introduced in the U.S., it was customary to ship the living animals to the eastern parts of the country.

Cattle Slaughtering

In recent years labour-saving devices have been widely adopted in the dressing and processing of cattle. A continuous rail system is generally used today, whereby cattle are chained by one leg and hoisted to a movable pulley on an overhead rail and are then slaughtered; from this position blood leaves the body so quickly that death occurs almost instantly, and for purposes of kosher ritual this method of slaughtering is defined as humane even without prior stunning.

The carcasses then move slowly along a continuously moving rail to stations where each required process is completed, including skinning, disemboweling, and beheading. Before going into refrigeration, the carcasses are cut down the backbone and split into sides.

Beef is shipped to wholesalers and retailers as wholesale cuts, such as chucks, rounds, and loins; in fast-distribution plants the carcass is reduced all the way down to retail consumer cuts. Sides and quarters of beef are still sold, but they constitute less than 5 percent of sales.

Hog Slaughtering

Traditionally, hog slaughtering has involved more complete processing at the packing plant than cattle slaughtering. The carcasses of slaughtered hogs also move on continuous rail or chain systems. Most often the hog carcass is first conveyed through scalding vats to dehairing machines. Hogs generally are not beheaded or dismembered in any way during the slaughtering and dressing process, which includes eviscerating, washing, and trimming.

Later, however, the carcasses are usually carved up into such cuts as loins, legs, and picnic hams or shoulders. Certain cuts, including loins, are sold fresh, without processing, but most cuts go through one or more processing operations. Although some plants still soak particular cuts in barrels filled with brine, it has generally been necessary to shorten the curing process. To prepare hams, a curing solution is usually pumped internally.

Hog bellies, which are used for bacon, are mechanically saturated with curing solution through hollow needles. Many cuts are thoroughly or lightly smoked at the packing plant. Nearly 70 percent of all hogs are slaughtered and fabricated into cuts at the packing plant and shipped to commercial processors who produce a complete line of sausage and cured products.

Legislation

Although inspection by the federal government of meat used in the packing industry was provided for in earlier acts of Congress (1890-91, 1895), comprehensive legislation was not introduced until 1906. By the act of 1906 all cattle, sheep, goats, and hogs became subject to antemortem and postmortem examination when the meat was to be used in interstate or foreign commerce; later the act was extended to include reindeer. By this legislation about 60 percent of the total meat supply of the U.S., was brought under inspection. The Packers and Stockyards Act of 1921 added further control, directed against trust activities. The Wholesome Meat Act of 1967 requires all meat-packers to meet federal inspection requirements, and agreements to that effect have been made between the various states and the federal government.

Recent Trends in Meat Packaging

Packaging is a scientific method of preserving the quality and safety of the food it contains from the time of manufacture to the time it is used by the consumer and also to display the product in the most attractive manner for consumer preference. The ability to convince a potential buyer to use a given product has always been influenced by product packaging and has become an important marketing tool for communicating brand values.

Besides protecting the product from physical, chemical, or biological damage, it should also enhance marketability of the product and be environment friendly. The only criterion customers have at the point of purchase of fresh meat cuts is visual appearance. Consequently, desirable meat colour must be maintained during chilled storage, distribution and subsequent retail display if preservative packaging systems are to be effective. In the processed meat products, other factors such as dehydration, lipid oxidation, discolouration and loss of aroma must be taken into account.

Packaging of meat and meat products has always been challenging because of perishable nature of the product. Today, packaging constitutes an important department of a food processing industry and many meat-packaging systems are available with different attributes and applications. These systems range from overwrap packaging for short-term chilled storage and retail display to 100 per cent carbon dioxide atmosphere packaging for long-term chilled storage.

Modern meat packaging technologies include vacuum packaging, modified atmosphere packaging, controlled atmosphere packaging, active packaging, smart packaging, etc. These strive to enhance the food safety and quality in as natural way as far as possible. The trends in innovative packaging techniques are discussed in the text that follows.

Packaging Forms and Materials

Packaging constitutes both flexible and rigid materials, alone or in combination with other preservation methods, to offer the necessary barrier, inactivation, and containment properties required for successful food packaging. Examples of typical flexible packaging materials include polyester, polypropylene, polyethylene, polyvinyl, polyolefins, vinylidene, vinyl chloride, surlyn, or nylon, which provide the needed characteristics such as strength, permeability, and saleability in the food industry.

Additionally, flexible packaging materials can be combined with gaseous environments to extend the shelf life of many foods, including meat and poultry. Besides synthetically derived packaging materials, flexible packaging can encompass the use of edible films, gels, or coatings made from polysaccharides, proteins, lipids, or composites of any or all three. The benefits of using edible films as packaging materials are threefold: these films may resist the migration of outer moisture into the food during storage; films may serve as gas and solute barriers; and films compliment other types of packaging by improving the quality and shelf life of foods.

And finally, the combination of rigid packaging materials made from metal, glass, or plastic and combined with heat, provides the most effective and widely used method of food preservation. The various traits of materials used in packaging of meat. Nearly all-available films for meat packaging are of thermoplastic materials, and therefore heat seal able, resulting in hermetically sealed plastic pouches, bags, etc. A high oxygen barrier is important in the application of films for packaging meat and meat products.

Films made of polyvinylchloride (PVC), polyethylene (PE) or polypropylene (PP) have a relatively high oxygen permeability, whereas polyvinylidene chloride (PVDC), polyester (PETP), polyamide (PA) and cellulose film (ZG) are less or almost non-permeable to oxygen. The latter materials are therefore better suited for packaging meat and meat products.

However, the materials of the first group are frequently used as laminates with materials of the second group in order to achieve special effects regarding mechanical strength, heat sealing properties or making the package practically impermeable to both oxygen and water vapour.

Packaging Techniques

Choice of packaging systems and the use depend on many

factors including cost, the volume of product being handled, and presentation. Some of the recent methods used in meat industry are:

Vacuum Packaging

Vacuum packaging refers to packaging in containers (rigid or flexible), from which substantially all air has been removed prior to final sealing of container. This method of packaging is actually a form of "modified atmosphere" since normal air is removed from the package.

The various advantages of vacuum packaging are that these extend shelf life by controlling oxidative rancidity and by preventing the growth of normal spoilage bacteria (aerobic organisms such as pseudomonas are suppressed and lactic acid bacteria are favoured which can grow to high numbers without causing spoilage), reduce freezer burn and loss of moisture at product surface, require minimum storage space and Leakers are easily detected (small puncture or pin hole in a vacuum pack is easily detected by looking for loose packages).

Today vacuum packaging is the most widely used method employed to extend the storage life of fresh meat. The air within the package must be evacuated to normal anoxic levels, *i.e.*, less than 500ppm to prevent browning due to residual oxygen. The exclusion of oxygen from meat surface immediately after carcass is broken into primals preserves the meat's potential to re-oxygenate soon after retail pack display.

The preservative effect of vacuum packaging is by inhibition of potent spoilage bacteria at normal meat pH under optimum vacuum packaged conditions. The microenvironment within the pack will determine the type of micro-flora, which develops. When meat is first vacuum packaged any residual oxygen remaining in the pack is consumed by meat and muscle pigments and CO_2 is produced as an end product of tissue and microbial respiration.

Under a good vacuum the package head space consists of <1 per cent O_2 and 10-20 per cent CO_2 which restricts the growth of aerobic microorganisms such as pseudomonas and favours desirable facultative anaerobic organisms such as lactobacillus but also undesirable Bronchothrix thermosphacta.

These slow growing bacteria cause less offensive types of spoilage even at higher bacterial numbers than other aerobic organisms and will eventually cause spoilage but only after many weeks of storage. If vacuum packaging procedures are followed correctly, the storage life of meat can be extended.

Care must be taken to keep initial bacterial load as low as possible by following strict hygienic procedures. The temperature must be maintained at or near 0-degree Celsius. High pH and DFD meat must be avoided as early spoilage may occur. High vacuum levels result in the most desirable muscle colour and fat appearance. Vacuum packaged meat of normal pH (<5.8) can be stored for 12 to 14 weeks at 0°C. However, vacuum packaging is considered unsuitable for red meats for retail display purposes since the oxygen-depleted atmosphere causes the meat in these packages to appear purplish in colour and therefore not acceptable to consumers. Vacuum packaging is ineffective for whole carcasses or cuts which prevent the packaging film being closely applied to all surfaces.

There are four basic ways; the first one involves the use of heat shrinking flexible packaging material around the primal cuts. When exposed to heat it shrinks and increases film thickness, improves mechanical resistance and reduces the drip loss. The second type involves the use of preformed plastic bags/ pouches made up of polyamide (PA) as the outer layer, while the inner core is polyethylene (PE). Polyamide provides barrier properties and physical strength while polyethylene provides sealing properties.

The third type provides the use of thermoforming trays in line from a base web. Here the product is placed into the tray and then a film web is allowed to cover the tray from a second reel of film followed by pack evaluation and vacuum sealing. The fourth type is vacuum skin packaging, in which the meat is placed in a rigid preformed tray or on the flat surface of a flexible base material followed by heat softening, evacuation and vacuum shrink sealing.

Gas Packaging

Gas packaging can be defined as the alteration of the proportional volumes of the gases, which comprise a normal atmosphere.

This type of packaging generally falls into two categories:

1. *Modified Atmosphere Packaging (MAP):* Refers to enclosing a product in some type of barrier and modifying the atmosphere either by drawing a vacuum or filling with a gas mix. During storage the level of these gases will change due to respiration of the product and permeability of the film.
2. *Controlled Atmosphere Packaging:* A system where gases are added or removed to maintain a desired balance is called controlled atmosphere packaging. In this case the bulk bin or the storage vessel is virtually impermeable.

Advantages of Gas Packaging

Advantages of gas packaging are that it allows meat to bloom (if the oxygenated red colour is desired, blend incorporated oxygen can be used to control colour), extends shelf life by controlling oxidation by displacing the oxygen with CO_2 and N_2 inhibits growth of normal aerobic spoilage organisms besides surface molds, prevents crushing of soft products as gas mixtures of CO_2 and N_2 equalise the pressure in the package and prevent package collapse and retains moisture but water can move from centre of product to surface resulting in staling.

In MAP systems each gas has a specific role to play in extending the shelf life and maintaining the appearance of packaged meat. The most commonly used gases for the packaging of meat is CO_2, N_2 and O_2 although other gases including CO, N_2O, Argon, SO_2 and Ozone have been tried to a limited extent. The most commonly used gas mixture for fresh red meat is high O_2, which has a minimum 60-70 per cent O_2 and 30-40 per cent CO_2.

Packaging in high oxygen extends the time for occurrence of microbiological spoilage and discolouration of meat. Most chilled poultry products are sold in prewrapped O_2 permeable film, which prevents moisture loss and the spread of contaminating microorganisms. However, where O_2 impermeable films are used, mainly for duck and turkey, the benefits are in extended shelf life.

There are also reports that high CO_2 atmosphere packaging of poultry extends the storage life up to three times that of storage in air. The modified atmosphere packaging for cured meat products has generated certain problems viz colour stability of cured meat products stored under illumination during retail display. Light exposure including intensity of light or illuminance, oxygen transmission rate of packaging material and residual level of O_2 have been found to effect the colour stability of cured meats.

Discolouration in controlled atmosphere packaging can be prevented only by exclusion of all O_2 from the package, which requires special evacuation equipment and totally gas impermeable packaging materials. If CO_2 is a major or sole component of the input atmosphere, then the quantity of added gas must be adjusted to assure that the intended atmosphere persists after dissolution of the gas into the product. For chilled meat, the most effective technology to date is the high CO_2 controlled atmosphere packaging system. This regime limits microbial deterioration through optimal storage temperatures (-1.5°C), high levels of CO_2, low residual O_2 (=0.05 per cent), and use of a gas impermeable films.

Active and Intelligent Packaging

An active or intelligent packaging is a packaging, which has an additional desired role that improves the quality of the food product as a supplement to pure packaging properties. These properties can either be integrated in the packaging material or separated from it. Intelligent packaging contains an external or internal indicator for the active product history and quality.

It has an inherent ability to gather information on its operating environment or history, to process that information in order to draw intelligent inferences from it and to act on those inferences by changing its characteristics in an advantageous manner. Active packaging has an extra function in addition to that of providing a protective barrier against external influence. It can control, and even react to, phenomena taking place inside the package.

It actively and constantly changes either package permeation properties or the concentration of various volatiles and gases in the package headspace during storage, or actively adds microbial, antioxidative or other quality improving agents, *e.g.*, flavour enhancing substances, via packaging material into food in small amounts during storage.

Examples of active packaging are those that exhibit preferential permeability; temperature compensating attributes; oxygen, carbon dioxide, alcohol, or ethylene scavenging systems; moisture absorbers; antimicrobials incorporated directly into the packaging matrix; controlled release of colourants, flavours, spices, minerals, antimicrobials, ethanol, or antioxidants; or the ability to absorb odors. Based on their way of functioning, active packaging are categorised into four groups viz scavengers, emitters, indicators and others.

The first three types can be evaluated as groups, because of the similarities in the intended function of the product but for other types (*e.g.* Susceptor materials) of active packaging, an individual evaluation will have to be done from case to case.

Active Packaging Scavengers

Active packaging scavengers consist of ingredients, which are intended to absorb, remove and then eliminate substances, such as oxygen, ethylene, moisture, or taint from the interior of a food package. The constituents of active packaging material and articles have an effect on the shelf life or the organoleptic properties of the food. The examples are the oxygen scavengers and ethylene scavengers, which are used for eliminating oxygen and ethylene respectively from the package resulting in an extended shelf life of the food. Commercial oxygen scavengers are in most cases based on iron powder, which, in the presence of oxygen, forms iron oxide. Non-metallic and organo-metallic compounds have also been developed.

Active Packaging Indicators

Various indicators, *viz* indicators for temperature, microbial spoilage, package integrity, physical shock, and product authenticity etc can give information on the quality of the food product directly, besides the package and its headspace gases, as well as on the storage conditions of the package. Some indicators do not need to interact with the product or the headspace, while others do. These indicators are often called intelligent packaging and certain concepts are already commercially available, and their uses seem to be increasing. New concepts of leak indicators and freshness indicators are patented, and it can be expected that new commercially available products will be available in the near future.

Antimicrobial Agents

Quite recently, the idea of incorporating antimicrobial agents directly into packaging films has been developed. Thus, the packaging material can serve as a source of releasing preservatives or antimicrobial agents, or even prevent the growth of microorganisms.

Flores and others (1997) reviewed the products and patents in the area of active packaging and identified antimicrobial packaging as one of the most promising versions of an Active Packaging system. Of the active packaging applications, the incorporation and/or slow release of antimicrobials is receiving considerable attention as a means of extending the bacterial lag phase, slowing the growth rate of microorganisms, prolonging shelf life, and maintaining food safety.

The general premise of antimicrobial packaging is a controlled migration of the compound to the food through diffusion or partitioning (Han, 2000), which not only allows for initial inhibition of undesirable microorganisms, but also retains residual activity over time.

Active Packaging Emitters

This group of active packaging contains, or produces, substances, which are meant to migrate into the food packaging headspace or into the food in order to obtain a technological effect in the atmosphere in the packaging or in the food itself as, *e.g.*, food additives, flavourings or biocides. In these cases, the consumer together with the food ingests the components.

Edible Films

Considerable research has been conducted to develop and apply bio-based polymers mainly from a variety of agricultural commodities and/or waste of food products to reduce the amount of synthetic polymer waste. Edible coatings and films offer a variety of advantages such as biodegradability, edibility, biocompatibility, aesthetic appearance, and barrier properties. The benefits of using edible films as packaging materials are threefold. Since the water activity (aw) is critical for microbial, chemical, and enzymatic activities, edible films may resist the migration of outer moisture into the food during storage.

They may serve as gas and solute barriers and compliment other types of packaging by improving the quality and shelf life of foods. With the incorporation of antimicrobials into the matrix, edible gels, films, or coatings can exhibit active packaging characteristics, which are attributes necessary for controlling undesirable microbes in muscle foods.

Antimicrobial compounds such as organic acids (acetic, propionic, benzoic, sorbic, lactic, lauric), potassium sorbate, bacteriocins (nisin, lacticin), grape seed extracts, spice extracts (thymol, p-cymene, cinnamaldehyde), thiosulfinates (allicin), enzymes (peroxidase, lysozyme), proteins (conalbumin), isothiocyanates (allylisothicyanate), antibiotics (imazalil), fungicides (benomyl), chelating agents (EDTA), metals (silver), or parabens (heptylparaben) can be added to edible films to reduce bacteria in solution, on culture media, or on a variety of muscle foods.

The incorporated antimicrobial agents must be edible food grade compounds since they have to be consumed along with the edible gels, films, or coatings.

Biopolymers

Research efforts since 1940s have been devoted to develop bio-based polymer films derived from natural sources such as plant and animal materials. The use of plant material to form films is an active research topic. The advantages of using biopolymers for food packaging include: reduced dependence on petroleum-based packaging, use of a renewable agricultural resource, the biopolymers can act as carriers to deliver shelf-life extenders such as antimicrobials or antioxidants, and biodegradability.

Since the forming of a film requires cross-linking of molecular units to impart strength and flexibility, proteins and carbohydrates are often the best candidates for biopolymer films. Proteins investigated include wheat gluten, corn zein, whey, pea protein, meat proteins, egg proteins, and soy.

Starches studied in the literature include alginate, polysaccharides, cellulose, carrageenans, and microbial polysaccharides. Pearce and Lavers reported using carrageenan to protect frozen poultry and Klose *et al.* incorporated an antioxidant into a gelatin coating to slow the development of oxidative rancidity in cut poultry meat prior to freezing.

Edible film packaging of meat act as a moisture barrier, oxygen barrier, texture modifier, breading adhesion aid, mold suppressor, bacterial inhibitor, physical protectant, oil barrier, antimicrobial carrier and antioxidant carrier. The colour stability of freshly cut beef surfaces has been reported to improve by exposing the meat surface to antioxidant-impregnated corn zein films.

The latest trends in MAP and overwrap packaging

Usually discussions about packaging technology center on new advancements that have arrived or will arrive soon. Features are unveiled, and their unique processing benefits checked off and dissected. Less often, but perhaps more important, the conversation takes a different direction, to how packaging truly aids retail merchandising and addresses things important to shoppers.

Looking at modified-atmosphere packaging and overwrap packaging, little change has occurred in recent times. Broadly speaking, overwrap packaging is still the most common format in retail meat cases. Of course, much of that use comes from in-store packaging. In fact, according to most surveys, its dominance was slowly trending downward until a few years back, when some retailers of significant size and influence decided that the in-store wrapped look for certain key beef cuts was an important selling point for shoppers.

Their switch from case-ready, modified-atmosphere packages to PVC overwrapped packages distributed in mother bags stabilised the decline. Perhaps it is a sign of the times, but modified-atmosphere packaging has not experienced much growth either over the past two or three years. An experienced observer of the packaging and retail industries senses that MAP may have seen a slight uptick, with some gains in ground beef, but otherwise it has been basically flat. It remains the third most-common packaging format in the meat case after overwrap and vacuum packaging. There does not appear to be much to get excited over here. But circling back to shoppers and their interests, MAP seems to soothe a couple of important pains. According to a packaging industry source, the longer overwrap packages are in distribution or in the display case, meat purge tends to work its way over the tray lip and under the package.

Shoppers avoid leaky packages and those that appear leaky. They do not like how they look, and they certainly do not want to handle them. The hermetic seals of modified-atmosphere packages, if done properly, keep the packages clean and dry. Cross-contamination concerns are eliminated for shoppers and retailers, alike, and fewer products must be marked down, reground or thrown out.

But overwrap packaging's decades-long inertia continues on because it works. It is what has always been done, and moving away from it at anything more than a measured pace means change. And change is hard.

Successful change requires commitment to more than just switching packages. It means implementing education programmes not only for shoppers, but for store personnel as well. The market is up to doing this. But realistically speaking, retailers have so many priorities to worry about, that if a package is not causing pain and it works, then change will happen when it does.

Use of Nanotechnology in Meat Packaging

The flexible packaging industry is growing rapidly. It is a \$38 billion global market. With the demand for flexible packaging growing at an average rate of 3.5 percent per year, flexible materials need to meet and exceed the high expectations of consumers and the stressors of the supply chain. Increasing competition between suppliers, along with government regulations, have resulted in innovations in films that enhance product and package performance, as well as address worldwide concerns with packaging waste.

The consumer demands meat, to remain fresh for long time, ease in handling, safe and healthy with environmental friendly packaging. Properties such as mechanical and heat resistance can be increased. Packaging materials that have improved temperature performance can be used for hot fill operations. Very thin films that can offer the advantages of flexibility and functionalities like being anti-counterfeit, anti-tamper and anti-microbial should be made. Self-heating feature can also be incorporated in the packaging material. Environment friendly, lightweight-packaging materials can be made for use in army rations. In future, with the aid of nanocomposites we may be able to modify plastic into a super barrier just as glass or metal.

Nanocomposite Technology

Nanocomposites, defined as polymers bonded with "nano" particles to produce materials with enhanced properties are recently gaining momentum in mainstream commercial packaging use. One such innovation is polymer nanocomposite technology, which holds the key to future advances in flexible packaging. "Nanocomposites appear capable of approaching the elusive goal of converting plastic into a super barrier - the equivalent of glass or metal-without upsetting regulators.".

Polymer nanocomposites are constructed by dispersing a filler material into nanoparticles that form flat platelets. These platelets are then distributed into a polymer matrix, creating multiple, parallel layers that force gases to flow through the polymer in a "tortuous path," forming complex barriers to gases and water vapour. The more is tortuosity present in a polymer structure, the higher is the barrier properties. The permeability coefficient of polymer films is determined using two factors: diffusion and solubility coefficients, *i.e.*, $P = DxS$.

Different types of fillers are utilised; the most common is a nanoclay material called montmorillonite - layered smectite clay. Clays in a natural state are hydrophilic, while polymers are hydrophobic.

To make the two compatible, the clays' polarity must be modified to be more "organic." One way to modify clay is by exchanging organic ammonium cations [positively charged ions] for inorganic cations from the clays' surface. Many different types of commercial plastics, flexible and rigid, are utilised for nanocomposite structures, including polypropylene, nylon, polyethylene terephthalate (PET) and polyethylene.

Chemical giant Bayer produces a transparent plastic film (called Durethan) containing nanoparticles of clay. The nanoparticles are dispersed throughout the plastic and are able to block oxygen, carbon dioxide and moisture from reaching fresh meats or other foods. The nanoclay also makes the plastic lighter, stronger and more heat-resistant.

Nylon nanocomposites, used as barrier layers for multilayer PET containers, prove to perform as much as two to three times better than the traditional ethylene vinyl alcohol barrier layers, since nylon has a 50°F higher melt temperature. The advantages of nanocomposite films are numerous, and the possibilities for application in the packaging industry are virtually endless. Because of the nanocomposite process's dispersion patterns, the platelets result in largely improved performance in the following properties:

- Gas, oxygen, water, etc., barriers
- High mechanical strength
- Thermal stability
- Chemical stability
- Recyclability
- Dimensional stability
- Heat-resistance

Good optical clarity (since particles are nano-size). By 2009, it is estimated that the flexible and rigid packaging industry will use 5 million lb of nanocomposite materials in the food and beverage industry. By 2011, consumption is estimated to be 100 million lb. Polymer nanocomposites are the future for the global packaging industry. Once production and materials costs decrease, companies will be using this technology to increase their product's stability and survivability through the supply chain to deliver higher quality to their customers while saving money. The advantages that nanocomposites offer far outweigh the costs and concerns, and with time, the technology will be further refined and processes more developed. Other types of nanofillers, allowing new nanocomposite structures with different, improved properties that will further advance nanocomposite are being developed for uses in many diverse packaging applications.

Antimicrobial and Active Packaging using Nanotechnology

Nano materials helps to keep products fresh for a longer period of time by using nano-sensors placed in meat production and distribution facilities, meat packaging or the meat itself which can detect all kinds of meat borne pathogens like E. coli, Campylobacter, Salmonella and Listeria spp by attaching themselves to the pathogens. Nanostructured film can prevent the invasion of bacteria and microorganisms and ensure the meat safety, a consultancy said. With embedded nanosensors in the packaging, consumers will be able to determine whether meat has gone bad or find out its nutrition content.

Anti-microbial packaging materials are very useful. It is hoped that with the advancement and growth in nano-science we can expect almost all packaging materials with anti-microbial agents incorporated in them. By placing Nano-sensors in packages, problems such as contamination, toxins or lapses in expiry dates can be detected.

Another window of opportunity for nano-particles is represented by meat safety applications. Latour *et al.* (2003) are investigating the ability of synthesised adhesin-specific nano-particles to irreversibly bind to targeted types of bacteria, inhibiting them from binding to and infecting their host. This research is aimed at reducing the infective capability of human meat borne entero-pathogens in poultry products, using two types of nano-particles.

One type is based on the self-assembly of organic polymer (*e.g.*, polystyrene), and the other on inorganic nano-particles functionalised with polysaccharides and polypeptides that promote the adhesion of the targeted bacterial cells. Understanding the interaction between contaminated surfaces and microorganisms allowed the design of materials that are resistant to bacterial adhesion. Researchers are already making efforts to develop a new generation of "self-cleaning" materials loaded with anti-microbial compounds that can be released under certain environmental conditions and kill the contaminant micro flora.

Baker and coworkers have used high-shear mixing of a lipid oil discontinuous phase with an aqueous continuous phase to develop nano-emulsions, which is an effective solution for disinfecting sensitive equipment safely. Nano-emulsions consist of oil droplets of 400-800 μm in diameter that are able to fuse with and subsequently disrupt the membrane of a variety of different pathogens, such as bacteria, spores, enveloped viruses, and fungal spores.

Kodak, best known for producing camera film, is using nanotech to develop antimicrobial packaging for meat products are commercially available since 2005. Kodak is also developing other 'active packaging,' which absorbs oxygen, thereby keeping meat fresh.

Electronic Tongue Technology and Smart packaging

Today, meat packaging and monitoring are a major focus of meat industry-related nanotech R&D. Packaging that incorporates nano materials can be "smart," which means that it can respond to environmental conditions or repair itself or alert a consumer to contamination and/or the presence of pathogens. According to industry analysts, the current US market for "active, controlled and smart" packaging for foods and beverages is an estimated $38 billion - and will surpass $54 billion by 2008.

Scientists at Kraft, as well as at Rutgers University and the University of Connecticut, are working on nano-particle films and other packaging with embedded sensors that will detect meat pathogens. Called "electronic tongue" technology, the sensors can detect substances in parts per trillion and would trigger a colour change in the packaging to alert the consumer if a meat has become contaminated or if it has begun to spoil.

While devices capable of detecting meat-borne pathogens could be useful in monitoring the meat supply, sensors and 'smart packaging' will not address the root problems inherent in industrial meat production that result in contaminated meats: these problems are faster meat (dis)assembly lines, increased mechanisation, a shrinking labour force of low-wage workers, fewer inspectors, the lack of corporate and government accountability and the great distances between meat producers, processors and consumers. Just as it has become the consumer's responsibility to make sure that the meat one is consuming has been cooked long enough to ensure that pathogens have been killed. Consumers will soon be expected to act as their own meat inspectors so that industry can continue to trim safety overhead costs and increase profits.

Nanotechnology and Transportation of Gases

Through nano-technology we can increase or decrease gas transmission rate to suit the packaging requirements of the product. Nanotechnology enables designers to alter the structure of the packaging materials on the molecular scale.

With different nanostructure, plastics can gain various gas and water vapour permeability to fit the requirements of preserving fruit, vegetable, beverage, wine and other foods. By adding nano particles, people can also produce bottles and packages with more light- and fire-resistance, stronger mechanical and thermal performance and less gas absorption.

Such nano tweaking can increase the shelf life of meats and preserve flavour and colour. SINTEF-Nanotech to create small particles in the film and improve transportation of some gases through the plastic film to pump out dirty air such as carbondioxide has been developed.

Dispersing a filler material into nano particles, creates multiple, parallel layer that force gases to flow through the polymer in a "tortous path", forming complex barriers to gases and water vapour

Nanotech Bio-switch in 'Release on Command' Meat Packaging

Researchers in the Netherlands are going one step further. They want to develop intelligent packaging that will release a preservative if the meat within begins to spoil. This "release on command" preservative packaging is operated by means of a bio-switch developed through nanotechnology to extend the shelf life of meat/ food. Research project in Finland is working on printable indicator, which contains reactive substance that signals if oxygen is present in package.

Meat Packaging Sensors in Defence and Security Applications

Developing small sensors to detect meat-borne pathogens will not just extend the reach of industrial agriculture and large-scale meat processing. The US military however views it a, it's a national security priority. With present technologies, testing for microbial meat-contamination takes two to seven days and the sensors that have been developed to date are too big to be transported easily.

Several groups of researchers in the US are developing biosensors that can detect pathogens quickly and easily, reasoning that "super sensors" would play a crucial role in the event of a terrorist attack on the food supply. Researchers at Purdue University are working to produce a hand-held sensor capable of detecting a specific bacterium instantaneously from any sample. They've created a start-up company called BioVitesse. Environmental friendly, lightweight-packaging materials can be made for use in army rations.

Overall Advantages of Nanotechnology on Meat Packaging

- Thermal Stability and Chemical Stability
- Heat resistance
- Gas, Oxygen, water, etc., barrier.
- Dimensional stability and Flexibility
- Functionality - Anti-counterfeit, Anti-tamper, Anti-microbial, Sensors (temperature, moisture, light, decay)
- High mechanical Strength
- Good optical density
- Recyclability.

Companies Working on Nano Meat Packaging

- *CONSTAR® International:* "Diamond Clear" - nanomaterial blended with polyethylene terephthalate - oxygen scavenger
- *Air product polymer:* "AIR FLEX EF 9100 EMULSION" - prevent penetration of liquid or gases and environment friendly.
- *KODAK®:* Antimicrobial food packaging
- Amcol International Co-op - NANOCOR®: Polymer blended with nanocrystal

- *Honey well®:* Polymer nanoclay - gas barrier
- *OX:* Oxygen scavenger
- *HFX:* Hot Filled Application
- *CSD:* Carbonated soft drink
- *SOLPHAS®:* Environment friendly nano-particle for coating plastic film, Antimicrobial packaging
- *BAYON®:* Supply nano-composite nylon film to juice carton
- *KRAFT®:* Working on nano-particle films that will detect meat pathogens

Global Overview of Nanotechnology Food

A handful of food and nutrition products containing invisible nano-scale additives are already commercially available. Hundreds of companies are conducting research and development (R&D) on the use of nanotech to engineer, process, package and deliver meat and nutrients to our shopping baskets and our plates. Among them are giant food and beverage corporations, as well as tiny nanotech start-ups.

The U.S., is leading in nanotechnology research, with more than 400 research centers and companies involved and more than $3.4 billion in funding. Europe has more than 175 companies and organisations involved in nanoscience research, with $1.7 billion in funding, and Japan has more than 100 companies working with nanotechnologies. Globally, the market for nanocomposites is expected to grow to $250 million by 2008, with annual growth rates projected to be 18 percent to 25 percent per year.

A 2004 report produced by Helmut Kaiser Consultancy, "Nanotechnology in Food and Food Processing Industry Worldwide," predicts that the nanofood market will surge from $2.6 billion today to $7 billion in 2006 and to $20.4 billion in 2010.

Status in India

To promote research and development in nano-science the Government of India has launched a programme "Nano-Material Science And Technology Initiative" (NSTI) in tenth five-year plan One hundred crore rupees have been allocated. Yet, this technology is in budding stage and not entered the field of meat science and technology. Mumbai based "Yash Nanotech" is a business information provider and consultant to entrepreneurs and is aiming to become leading suppliers of nanotech tools, products and services in India.

Hazards and Risks

Although the hazards and risks of nano-technology are not known but can be assumed that like any other technology this may also have associated hazards and risks. Environmentalists are afraid that nano-technology may produce contaminants, which because of their nano-size, may pose to be ultra-hazardous. Even if these particles are not harmful, their interaction with products may be harmful.

However, judicious use of nano-technology will find tremendous applications in various fields. Research has shown that nano-sized particles accumulate in the nasal cavities, lungs and brains of rats, and that carbon nano-materials known as 'buckyballs' induce brain damage in fish. Vyvyan Howard, a toxicologist at the University of Liverpool in the United Kingdom, has warned that the small size of nanoparticles could render them toxic, and is of the opinion that full hazard assessments are needed before manufacturing is licensed.

Many interested parties, including the Canadian ETC Group and the insurance company Swiss Re, have expressed their concern over releasing tiny particles, which, because of their small size, are able to travel very far into the environment. They warn that we do not yet know how these particles will act in the environment or what chemical reactions they will trigger on meeting other particles.

However, these same groups also concur with nanotechnology advocates who feel the field may offer 'cleaner' technologies, and, ultimately, a cleaner environment. But mostly, the concern is for the lack of research into nanotechnology's potential threats to human health, society and the environment. After witnessing widespread rejection of genetically modified meats, the meat industry may be especially skittish about owning up to R&D on "atomically modified" meat products.

Application of Nanotechnology as a Tool in Animal Products Processing and Marketing

Nanotechnology and Livestock Products

Nanotechnology is quite competent in new products and new processes development with the objective of enhancing the performance of the products, extending the product shelf life and freshness as well as in getting better the safety and quality of food. Nanotechnology is an enabling technology that has the potential to modernise the food industry. The major challenges in improving nano based technologies are integration of the hardware (nanomachines), software and management tools for generation of new information's on this aspect.

Nanotechnology can also reduce the time of production of eggs and meat. Thus, consumers can get faster eggs and faster meat. The supply chain for these products may also be managed as an engineering industry. The time reduction in production of livestock and poultry are due to radical intervention of oviduct physiology at nano level and the cutting edge genetic reengineering.

The new nanoimaging techniques, new quantitative analytical tools, quantitative integration of information by using the microchips and the nanoprobes will alter the frontiers of the livestock and poultry physiological research. The qualities of fermented livestock products are now well proven for health benefits which can easily be obtained by this vital technology.

The development of meat industry may also depend on an ability to track every stage in the life of the product, including the birth of the animal, its medical history and its movements between the farms, the slaughterhouse and the meat-packing plant, right through to the consumer's table.

Nano Meat Production

Nanotechnology is a promising technology that has applications in almost all fields. The technology has opened the path to an unexplored science for studying individual nanoparticles and their unique application for poultry and meat industry ranging from meat design, achieving food security, meat safety, overcoming food allergies, eliminating pesticide use, meat packaging, restoring meat damage and sensory evaluation to processes such as filtration, separation, encapsulation, etc. Nano-technology can make poultry and meat products cost-effective with the natural properties. Production may be carried out by self replicating nano-devices using small assent of material, energy, low capacity, less labour and land. Thus, production is more efficient. For quality meat products development nanotechnology based diagnostic techniques may replace the ultrasound like existing techniques.

Nanotechnology already exists naturally in foods and meat is naturally composed of nanofibres. These nanofibres undergo changes during cooking or processing, which in turn influence the texture and eating quality. For manufacturers to deliver a successful vegetarian alternative that gives the taste and texture of meat, they need to understand and control the assembly of structures at the nano level.

The differentiation in qualities of the products can easily be made by nanotechnology based techniques. Designer poultry meat has to be designed in a cost effective way that strikes a rebut balance between sensory characteristics. One of the more futuristic applications of nanotechnology lies in the production of "interactive" poultry meat that change colour, flavour or nutrients depending on diner's taste or health. Many of the molecular structures that determine these characteristics are in the nanometer range and information on the source can play an important role in the poultry meat design.

The purpose is to master over the characteristic of meat components in an intelligent manner by manipulating atoms individually and place them exactly where they are needed to produce the desired flavour, texture, etc.. The potential of nano-technology in poultry meat industry cannot be fully appreciated yet because of lack of sufficient knowledge. If Nanotechnology continues to advance at its current pace, we could expect that soon we will be able to create unlimited amount of meat by synthesis at the atomic level, which would eradicate hunger.

Nanotechnology has the potential to revolutionise the global food system. Novel agricultural and food security systems, disease treatment delivery methods, tools for molecular and cellular biology sensors for pathogen detection, environmental protection and education of the public and future workforce are the examples of the important impact that nanotechnology could have on the science and engineering of agriculture and food systems. There were many methods to improve livestock meat products by nanotechnology.

Recently, nanotechnology and vegetarian meat, the optical tweezer ability to move individual particles around has intrigued nanotechnologists, who have wildly imaginative plans for what to do with the molecular scale sized robots they would like to create but so far, having few tools with which to make them. Nanotechnology holds out enormous possibilities and although, so, far relatively little has been accomplished, tons of money is being poured into the research, suggesting as nothing else that it is taken seriously.

The holy grail of nanotechnology is some version of an 'assembler', a robot the size of a molecule that would allow moving matter at the atomic and molecular level. The obvious power of such a technology given that everything is made of the same basic atoms but simply arranged in different ways is that we would be able to construct virtually any substance we wanted from scratch by putting together exactly the molecules we wanted.

Interestingly, one of the first examples given of the speculative technology of nanotechnology was that of synthesised meat. In short then, technologies ranging from the actual to the speculative promise a variety of ways to create real meat without killing animals. Although, still commercially infeasible at the moment or in some cases technically infeasible for several years to come, the point here is not to be unfocused by the fact that we cannot yet make exploit of these technologies but rather to choose whether we should support the development of these technologies. Some of the researchers in this field, for instance, are so committed to the development of cultured meat; largely out organisations to pursue the technology. For example, New Harvest is a 'non-profit research organisation working to develop new meat substitutes, including cultured meat-meat produced *in vitro*, in a cell culture, rather than from an animal'.

Cultured meat has the possibility to make eating animals unnecessary, even while fulfilling all the nutritional and hedonic requirements of meat eaters. It also has the potential to greatly diminish animal suffering. As such, the development of cultured meat would seem to have a moral claim on us; whether moral vegetarians for whom a greater opportunity exists to reduce animal suffering or conflicted meat eaters for whom practice could now cohere with beliefs or even for recreational hunters for whom ancillary arguments about providing food would fall by the wayside and require defences of getting pleasure from animal death *per se*.

The development of cultured meat, then, is not merely an interesting technological phenomenon, but something that we may be morally required to support. In doing so, we recognise that morality is not something that must simply respond to new technologies as they arrive, throwing us into confusion, but rather that morality may champion and assist in the development of new technologies, as a step towards the production of a world that in fact and not merely in ideal, mirrors the moral vision we possess for it.

Nano Milk Production

Nanotechnology is a new technological tool in modern raw milk production and pasteurisation. Recent and ongoing advances in biomedical technology may assist in advancing our understanding of disease prevention and health promotion, as well as medical diagnostics and therapeutics. Likewise, these advances will also soon open the barn door and accelerate our understanding of identical aspects in relation to the milking cow considering the ongoing progress within livestock genomics.

These new, emerging technologies such as microarray technology and nanotechnology have the potential to advance nutrition and health science in many aspects of relevance for modern milk production. This implies both a better understanding of the aspects of importance for ensuring the establishment of the conditions for the "whole cow" and hereby the necessary public acceptance of an effective and industrialised milk production and the support of milk and dairy products as a natural part of a healthy diet.

The simultaneous development within information technology makes it already now possible to integrate all data obtained throughout the milk chain. Consequently, data obtained through existing devices in the milk chain in combination with new data coming from the future devices for data collection based on the above mentioned new emerging technologies would be a unique possibility to develop solid decision support systems at all levels in the future.

Moreover, nanotechnology is meant to be a major driver in the:

- Development of biosensors
- Support of sustainable agriculture,
- Pathogen and contaminant detection,
- Improved animal health, through development of smart treatment delivery systems and
- Material science and engineering development of material with antibacterial activity and which is easy to clean. These are all areas that can benefit from a modern raw milk production.

Nano Egg Production

Role of nanotechnology in designer eggs production is now a well known fact. In future the share of designer eggs in the total share of the egg market will rise to more than 30 per cent in 2020. It is expected that by 2000, market has to supply the cholesterol free eggs, yolkless or reduced yolk eggs which can be the high value protein source, immune eggs which can supply the predetermined antibodies and therapeutic eggs with supply the predetermined physiological factors for treatment purposes. The tools and techniques currently with us will not give the solution for these challenges. They can only be meet out by the emerging nanotechnology, which deals not merely at the molecular level but at the atomic level.

Furthermore we can know in advance about the source of pathogens in eggs and poultry meat. Early detection of food borne pathogenic bacteria is critical to prevent disease outbreaks and safeguard public health. Numerous methods have been developed in order to detect this pathogen; however, the biggest challenges remain detection speed and sensitivity. Now, a novel nanotechnology-based biosensor is showing great potential for food borne pathogenic bacteria detection with high precision.

Nanotechnology Based Meat Packaging

The flexible packaging industry is growing much quickly and having $38 billion market worldwide. The demand of flexible packaging is increasing at an average rate of 3.5 per cent per year. Increasing competition between suppliers and government regulations, have resulted in innovations in films that enhance products and package performance, as well as address worldwide concerns with packaging waste.

The consumer demands meat, to remain fresh for longer period, ease in handling, safe and healthy with environmental friendly packaging. Properties such as mechanical and heat resistance can be augmented by the use of nanotechnology. Packaging materials that have improved temperature performance can be used for hot fill operations.

Very thin films that can offer the advantages of flexibility and functionalities like being anti-counterfeit, anti-tamper and anti-microbial should be made. Self-heating feature can also be incorporated in the packaging material. Environment friendly, lightweight-packaging materials can be made for use in army rations. In future, with the aid of nano-composites we may be able to modify plastic into a super barrier just as glass or metal.

Nanotechnology provides food scientists with a quantity of ways to make novel laminate films suitable for use in the food industry. A nano-laminate consists of two or more layers of material with nanometer size that are physically or chemically bonded to each other. Nano-laminates can provide food scientists some advantages for the preparation of edible coatings and films over usual technologies and may thus, have a number of significant applications within the food industry.

Edible coatings and films are presently used on an ample type of foods, including fruits, vegetables, meats, chocolate, candies, bakery products and French fries. These coatings or films could serve as moisture, lipid and gas barriers. Alternatively, they could improve the textural properties of foods or serve as carriers of functional agents such as colours, flavours, antioxidants, nutrients and antimicrobials.

The basic functional properties of edible coatings and films depend on the characteristics of the film-forming materials used for their preparation. The composition, thickness, structure and properties of the multilayered laminate formed around the object could be controlled in a number of ways, including changing of the type of adsorbing substances in the dipping solutions, the total number of dipping steps used, the order that the object is introduced into the various dipping solutions, the solution and environmental conditions used (pH, ionic strength, dielectric constant, temperature, etc).

The dynamic force for adsorption of a substance to a surface would depend also on the nature of the surface and the nature of the adsorbing substance and it could be: Electrostatic, hydrogen bonding, hydrophobic interactive, thermodynamically incompatible, etc.

The fresher force: extend shelf-lives

As consumer demand continues to shift towards the ready availability of fresh produce, the packaging industry is working to extend the shelf-life of fresh food with a range of innovative packaging concepts. Here we explore some of the industry's most promising life-extending technologies, from outline concepts still deep in development stages to products on store shelves today.

CO Gas-Flushing for Fresh-Looking Red Meat

"By adding a small plaster-style strip at the bottom of strawberry punnets, the ethylene gets absorbed and the berries stay fresh." Meat technologists and microbiologists have known for many years that gas mixtures containing carbon dioxide slow down bacteria such as pathogens, which could cause potential diseases, and pseudomonas, which are responsible for the development of slime and odours on meat stored in the fridge. Modified atmosphere packaging (gas-flushing) has however increasingly been used to make meat look redder for longer to appeal to shoppers and consumers.

In the process, the amount of oxygen (O_2) is lowered in the package from 20.9 per cent to 0 per cent to slow down the growth of aerobic organisms and the speed of oxidation reactions. The removed oxygen is then replaced with carbon monoxide (CO), which keeps the colour of the meat red.

Even though there was no risk found in the use of low levels of CO, some critics and consumer groups, as well as the

European Food Information Council (EFIC), are concerned that it can hide visual evidence of spoilage, asking for special labels that indicate when red meat has been gas-flushed. However, microbiologists widely recognise the process as a safe form to prolong the shelf life of red meat and make it more appealing for consumers.

Active Packaging: Antimicrobial Film

The development of active packaging, which upgrades packaging from a passive barrier to a proactive method of protecting food and extending shelf-life, has been building momentum in the industry over the last decade. Keeping organic foods fresh for longer has positive implications for both consumer safety and in ongoing efforts to reduce food waste. Given the potentially revolutionary effect of active packaging on the shelf-life of products like meat, fish, cheese and fresh vegetables, it's unsurprising that research in this area is receiving funding on a governmental level.

Indeed, the EU's 7th Framework Programme for Research and Technological Development has provided funding for one of the latest advances in active packaging. The Fraunhofer Institute of Process Engineering and Packaging IVV has developed an antimicrobial film that releases an agent on to the surface of the food it is wrapped around to eliminate bacteria and other pathogens which accumulate over time.

"The surface is the primary point of attack for germs," said Fraunhofer IVV food chemist Carolin Hauser. "Using only the smallest quantities of active agent, the packaging thus provides effective protection for food."

Fraunhofer chose sorbic acid for its active antimicrobial agent, as it is non-toxic and non-allergenic, as well as being neutral in taste and smell. In 2010 tests, the antimicrobial film destroyed around three quarters of the E. coli pathogens that had been injected into pork loin samples.

M&S Breaks Mould with Ethylene Absorber

Red, sweet and tasty - that is how consumers like their strawberries. But all too often the temptation turns into a rotten nightmare. "This 'killer paper' has been shown to wipe out bacteria like E. coli and S. aureus in just three hours." To blame is ethylene, the hormone that causes fruit to ripen and then turn mouldy. In order to slow down this process and extend the fresh look of strawberries, British supermarket giant Marks & Spencer (M&S) introduced a simple but ground-breaking new packaging in January 2012.

By adding a small plaster-style strip at the bottom of strawberry punnets, which contains a mixture of clay and other minerals, the ethylene gets absorbed and the berries stay fresh. The 8cm × 4.5cm strip, manufactured by the British firm It's Fresh, prolongs the shelf life of strawberries of up to two days without affecting the recyclability of the packaging.

M&S agronomist Hugh Mowat said: "This new technology is a win-win for our customers - not only will their strawberries taste better for longer, but we really hope it will help them to reduce their food waste as they no longer need to worry about eating their strawberries as soon as they buy them."

The retailer claims that previously carried out trials have shown a minimum waste saving of 4 per cent, equating to 40,000 packs or 800,000 strawberries a week during peak season. Should it prove successful, the retailer plans to introduce it for all its berries.

Heat-Resistant Chocolate Packaging

In 2010, when Kraft Food started looking for a new hi-tech thin film for packaging its chocolate bars, the food giant marched out to tackle one of the biggest issues the confectionary industry still has: as tempting as chocolate bars are, they quickly become very unappealing when they start melting at temperatures of around 33°C and get a white 'bloom' discolouration when stored in direct sunlight.

No breakthrough has been reached so far, but the area of printed electronics is opening up attractive new potential. One idea is to provide energy from printed batteries or solar cells, which could trigger a cooling process once the packaging gets too warm.

Photovoltaic (PV) company Kornaka has already introduced solar cells, which use a wider spectrum of visible and invisible light sources to generate power. The photo-reactive materials can be printed or coated onto flexible substrates using roll-to-roll manufacturing, similar to how newspaper is printed. Made from semi-conducting polymers and nano-engineered material, the printed active fabric is squeezed between printed electrodes, the substrate and the packaging material, and absorbs photons to trigger the release of electrons which are then transported to create electricity. The finished material is only about 50μm-250μm thick and could therefore be just right for the use for chocolate packaging.

Nanotechnology: Silver Nanoparticles

"Modified atmosphere packaging (gas-flushing) has increasingly been used to make meat look redder for longer." Nanotechnology might operate on the tiniest scale possible, but it has massive potential in a range of fields, not least that of packaging.

We've already looked at the implications of the 'nano-brick' concept being developed at Texas A&M University, which is claimed to be 100 times more oxygen-impermeable than standard plastic bottles by researchers. Another nano-concept catching the packaging world's attention is the use of silver nanoparticles. The germicidal properties of silver compounds have been well known for centuries, but new research suggests that tiny silver nanoparticles, no wider than 1/50,000 the width of human hair strand, can be applied as an effective antimicrobial coating on to packaging materials like plastic and metal.

Researchers at Bar-Ilan University in Ramat Gan, Israel, reported to the journal of the American Chemical Society that they had used ultrasound techniques to attach silver nanoparticles to a paper surface. This 'killer paper', as it has been dubbed, has been shown to wipe out bacteria like E. coli and S. aureus in just three hours.

Using nanoparticles, the researchers have been able to extend antibacterial potency beyond what is achievable with larger silver particles. The major challenge that remains for the academic world is to find a practical way of refining the technique so that it's a practical preservative packaging option for supermarkets and other food outlets around the world.

The Sustainable Germ Killer

As well as improving consumer safety and food waste,

another major priority for the development of intelligent food packaging is the reduction of packaging weight while maintaining recyclability. To this end, a pan-European R&D project called Wheylayer has been developing a method of coating plastic film in a biomaterial made from natural whey protein rather than petrochemical-based polymers like ethylene vinyl alcohol (EVOH).

The latest results from this research have revealed this substance's potential as an efficient, sustainable and economically viable alternative to petrochemicals. The whey protein-based layer is completely biodegradable and forms an effective shield against water vapour and oxygen, while the correct concentrations of bio-based softeners and additives allow the whey proteins to survive the mechanical loads involved in the manufacturing process.

Perhaps most promising is the fact the Wheylayer researchers have developed an industrial-scale method of applying whey protein layers to film and combining them with other intelligent films to increase protection. "We've managed to develop a whey protein formulation that can be used as the raw material for a film barrier layer," said Markus Schmid of Fraunhofer IVV, one of the partners working on Wheylayer.

"We have also developed an economically viable process which can be used to produce the multifunctional films on an industrial scale." If the project's production method is as simple and effective as scientists believe, Wheylayer could be well placed to meet increasing European demand for sustainable composite films in the near future.

Impact of Packaging Methods on Meat Preservation

In most developing countries meat carcasses are handled and stored either without or with only minimal refrigeration. A fast-turnover system is used, ensuring that meat slaughtered during the night and morning is sold and consumed the same day. This well-proven traditional method functions satisfactorily because in the short period between slaughtering and consumption, micro-organisms cannot increase to the extent of spoiling the meat and making it inedible. This system is eminently suitable for rural villages and small towns in many developing countries, particularly because local cooking habits do not call for ripened and tender meat. This fast turnover of meat normally does not lead to meat spoilage, although massive contamination during slaughtering and meat handling frequently occurs.

Fortunately, in most cases toxic micro-organisms do not develop to such an extent to threaten consumers' health, provided that the time between slaughter and meat cooking is short and cooking is done properly and with sufficiently high temperatures.

In traditional meat handling, fresh meat is generally not packaged at all or just wrapped in paper, leaves, etc. Meat, traditionally preserved by drying, is sometimes packaged in linen bags, baskets or pottery to facilitate storage and transport and to provide some kind of protection against dirt, insects, etc. With higher concentrations of population, however, this traditional system now becomes outmoded in some places in developing countries because more time is needed between slaughtering and ultimate consumption.

Meat frequently has to be stored, transported, prepared and distributed through a retailer or supermarket, all of which is considerably time-consuming. In order to safeguard fresh meat during this extended time, certain methods of preservation have to be applied. Refrigeration is the obvious solution, but this is expensive and therefore frequently not available in developing countries.

Energy-saving storage methods are therefore particularly relevant in underdeveloped areas. For both methods, either using the refrigeration technology or energy-saving methods to extend the shelf-life of meat and meat products, proper packaging has an important part to play.

Purpose of Packaging Fresh and Processed Meat

The purpose of packaging is primarily to protect foodstuffs during the distribution process, including storage and transport, from contamination by dirt, micro-organisms, moulds, yeasts, parasites, toxic substances or those influences affecting smell and taste or causing loss of moisture. Packaging should help to prevent spoilage, weight losses and enhance customer acceptability. Simple packaging without further treatments is less effective in prolonging the shelf-life of meat and meat products. Frequently full advantage of packaging can only be achieved in combination with preservation methods.

Factors Affecting the Shelf-Life of Meat and Meat Products

Though meat handling, storage and consumption may differ from one place to another, the factors limiting the shelf-life of these products are the same.

There are endogenous factors, such as:

- pH-value or the degree of acidity of the product;
- A_w value or the amount of moisture available in the product; and exogenous factors, such as:
- Oxygen (from the air);
- Micro-organisms;
- Temperature;
- Light; and
- Evaporation and desiccation.

Combined Effect of PH- and Aw-Value

Generally speaking the shelf-life of meat and meat products will be longer the lower the pH-value and/or a_w-value. Both factors (either pH or a_w alone or the two together) have a decisive influence on the growth of micro-organisms in food. However, there are limits for most meat products regarding decreased pH-value and a_w-value, particularly for organoleptic reasons. Except for some special products, consumers do not want meat products to be excessively acidic or dry.

Uncanned meat products can be classified into three storage groups according to their pH and a_w. Each group requires different storage conditions. Highly perishable meat products have a pH-value above 5.2 and an a_w-value above 0.95; refrigeration at or below +5°C is needed. These are raw fresh meat (without additives), bologna-type sausages, cooked sausages and cooked ham.

Perishable meat products have a pH-value below 5.2 or an a_w-value below 0.95. Refrigeration at or below +10°C is needed to keep them stable. Products such as meat or poultry pieces in vinegar jelly (acid) and semi-dry sausages or hams belong to this group.

Shelf-stable products have a pH-value of or below 5.2 and an a_w-value of or below 0.95, or only a pH-value below 5.0, or only an a_w-value below 0.91. No refrigeration is required in these cases, the products remaining stable under ambient temperatures. The most common products in this group are the various kinds of dried meat.

Under the above conditions no microbial growth in meat and meat products will occur. However, this does not mean that the products remain stable for an undetermined period. Their shelf-life will be limited by chemical or physical deterioration, by rancidity and discolouration. In this situation the product quality will benefit from the application of suitable packaging materials, which reduce the physical and chemical influences on the product or protect the product completely. The following noxious influences may occur.

Oxygen

The oxygen content in the air is about 20 percent. If oxygen affects meat and meat products during prolonged storage periods, it will change the red colour into grey or green and cause oxidation and rancidity of fats with undesirable off-flavours. The foils used for food packaging differ in their permeability to oxygen. The lower the oxygen permeability of the packaging material, the more efficient will be the protection of product quality. The best protection will be achieved using oxygen-proof packaging films together with vacuum packaging of the product. This ensures that practically no oxygen is left in the package and no oxygen will penetrate from the air into the product.

Light

The prolonged exposure of meat and meat products to daylight or artificial light accelerates oxidation and rancidity because light provides the energy for these processes.

Transparent packaging films give no protection against light influences. Therefore, for products under strong light exposure, coloured or opaque films should be preferred. Films laminated with aluminium foil are absolutely impermeable to light. Products in transparent packaging film are sufficiently protected when kept in the dark or under moderate illumination.

Evaporation

Fresh foods with a relatively high moisture content such as meat, fresh sausages, cooked ham, etc., will have considerable losses of weight and quality by evaporation during storage if they are not packed. The packaging material must therefore be sufficiently vapour-proof. Most plastic films used for food packaging comply with this requirement.

Secondary Contamination

During slaughtering, carcass dressing, meat cutting and/or processing, the contamination of meat to some extent cannot be avoided. The further growth of micro-organisms in meat and meat products cannot be stopped through packaging only. However, secondary contamination of these foods, for example by contact with dust, dirty surfaces and hands, can definitely be prevented through proper packaging, preferably with plastic films which are absolutely impervious to agents causing secondary contamination.

Suitable Materials and Equipment for Packaging Meat and Meat Products

Packaging films can be subdivided into cellulose films, plastic films and aluminium foil. They can either be used as monofilms or as two or more different films laminated together.

These materials differ in:

- Oxygen permeability;
- Water vapour barrier;
- Resistance to hot and cold temperatures; and
- Mechanical strength.

Nearly all available films are of thermoplastic materials, and therefore heat sealable, resulting in hermetically sealed plastic pouches, bags, etc.

A high oxygen barrier is important in the application of films for packaging meat and meat products. Films made of polyvinylchloride (PVC), polyethylene (PE) or polypropylene (PP) have a relatively high oxygen permeability, whereas polyvinylidenchloride (PVDC), polyester (PETP), polyamide (PA) and cellulose film (ZG) are less or almost non-permeable to oxygen.

The latter materials are therefore better suited for packaging meat and meat products. However, the materials of the first group are frequently used as laminates with materials of the second group in order to achieve special effects regarding mechanical strength, heat sealing properties or making the package practically impermeable to both oxygen and water vapour.

For the efficient utilisation of these materials, air must be drawn out completely from the packages with the meat product ("vacuum-packaging") and the package should be hermetically closed by heat sealing or metal clip. Vacuum-packaging will inhibit the growth of yeasts, moulds, and most aerobic bacteria under refrigeration temperatures. However, facultative anaerobes such as acid-producing bacteria grow freely under adequate aw-conditions.

These micro-organisms do not threaten the health of consumers but may affect the quality of the products by negatively influencing colour and taste. Therefore efforts should be made to keep the initial number of bacteria low, which can be obtained by simple hygienic measures during cutting and packaging, such as frequent handwashing, clean tools and clean tables.

The above general considerations refer to vacuum-packaged meat and meat products with a high moisture content. The situation is different for the traditionally preserved meat and meat products. All these products are shelf-stable because of their low water activity and therefore do not require refrigeration. They are partly dried in the sun, or others are dried and smoked over a wood fire.

These low-moisture products are hygroscopic and should be protected in the case of high air humidity. Traditional wrapping methods using paper, jute or linen will not inhibit the penetration of water vapour. However, a simple polyethylene bag is sufficient as a water vapour barrier, but care must be taken that the bag is tightly closed. If plastic bags are used for packing, then the pieces of dry meat should be cut to a certain length so that they can be better arranged in the bags. The bag should be bound with rope, rubber or cellotape. Bags stuffed with meat in this way are best stacked into a carton.

References

Fraser, Evan D. G.,Rimas, Andrew: *Empires of Food*, Random House, Delhi, 2011.

Fred Searing.: *Profit Building Ideas for Food and Food Service,* Dahl Publishing. Co., Stamford Conn, 1947.

G Chengappa, N Nagaraj and Ramesh Kanwar: *Challenges to Sustainable Agri-Food Systems,* I.K. International Publications, Delhi, 2007.

6

Modern Packaging for Fresh Poultry

Introduction

Poultry processors continue to work on developing innovative ways of packaging raw and cooked poultry products. Today's consumers are looking for convenience when it comes to storing, opening and cooking the product. Manufacturers are looking for packaging that address shelf-life; packaging that is flexible and versatile amongst multiple product platforms; and packaging that could potentially help processors, retailers and consumers continue to keep food safe at the plant, in the store and at home.

This spring, meat and poultry industry suppliers, manufacturers and processors will gather in Frankfurt, Germany, May 4-9, for the international IFFA show. IFFA is the leading international trade fair for packaging, processing and sales in the meat industry, held every three years. IFFA offers a comprehensive overview of the new products, ideas and trends in the packaging industry. At the 2013 show, Hall 11 of the Frankfurt Fair and Exhibition Centre will bring together the leading manufacturers of packaging technology, measuring and weighing equipment as well as top suppliers from the processing and cutting segment.

IFFA organisers expect to see "products showcased by the exhibitors that reflect the major challenges facing the entire meat industry." Show organisers said Hall 11 will feature, "products from packaging manufacturers designed for energy-efficient processing and further developments from the manufacturers of packaging materials in terms of barrier packaging."

The meat and poultry industry is looking for new packaging solutions utilising materials that are lighter, more transparent, more resistant during processing and also more flexible. "There are already solutions available on the market that optimise existing machinery with regard to energy efficiency and flexibility," according to IFFA representatives.

"The packaging sector is working at introducing products that are safer and have a longer shelf-life." Many of the packaging-materials manufacturers heading to IFFA produce new film-based concepts for new kinds of packaging for meat, poultry and other food products. These offer both longer shelf-life and more consumer-friendly handling (*i.e.*, ease of opening and resealability). Packaging companies will also address the continuing need to provide products that help with the marketing challenges of making poultry products more appealing to the customer at point-of-sale.

Packaging of Poultry Plant

A poultry plant is a facility where poultry are processed from slaughter to packaging for sale. Poultry are generally defined as domesticated birds such as chickens, ducks, turkeys, and geese. Poultry plants vary in size and scope, and some offer specialty services such as kosher butchering. Most are located in centralised areas with access to train and trucking lines which facilitate rapid transport of live birds and packaged meats, and some are attached directly to poultry farms.

At a poultry plant, live birds are slaughtered and processed to remove entrails and feathers so that they can be butchered for sale. Whole birds are often sold as-is, and the poultry plant can also provide processing for poultry parts such as legs, thighs, and wings. Scrap materials which are not prepared for sale are sent to rendering facilities, while the finished processed meat can be shipped in bulk to wholesalers, or packaged for shipment to grocery stores.

Specialty services which may be offered by a poultry plant can include kosher and halal meat processing, organic meat processing, and humane slaughter facilities. Meat from such plants tends to be more expensive, reflecting the increased care which must be taken in processing, and such plants can also be subject to more rigourous inspection. Government inspectors in these cases are concerned not just with food safety, but with evidence that the plant is conforming with regulations which allow it to sell meats labeled as humane, kosher, or organic.

Large scale poultry farms may have an attached processing plant to streamline their business operations. Not having to truck live birds cuts down on production costs, including costs associated with losing birds in transit.

Smaller scale farms may not want to run a processing plant, or may be part of a cooperative which pools resources to fund a centralised plant which handles birds from multiple farms. Very small farms may make arrangements for mobilepoultry processing, in which a slaughter truck comes to the farm to handle slaughtering and processing. Numerous workers are required for poultry plants to handle various stages of the operation.

The work tends to be difficult, loud, and dirty, and rates of compensation are usually fairly low because the labour is primarily unskilled. Working in a poultry plant also comes with a number of occupational hazards, thanks to the heavy equipment which is utilised in poultry processing. Plant workers can lose extremities to blades, conveyer belts, and other equipment, and they are also at risk of burns from scalding tanks used to remove feathers.

Europeans love roasted chicken, chicken legs and other tasteful poultry products. Almost three quarter of all poultry packs in Europe are sold at retail, with discount stores taking up an ever bigger portion. To increase market share against lower priced pig meat, which is still the number one meat product in volume, first-class quality is crucial. This requires that the particularly sensitive poultry is packed in a secure and hygienic manner.

"This requires that the particularly sensitive poultry is packed in a secure and hygienic manner!"The packaging industry provides a series of concepts for poultry products with highest possible protection and maximum shelf-life. Those who are looking for a low-cost solution usually choose flexible packaging.

This solution has long since become a standard in the poultry industry. The pack is made on a thermoformer from two flexible films and is sealed under vacuum. In the meantime, the market has moved forward. With modern technologies and packaging materials, shelf-life can be extended, retail optics improved and branding possibilities for manufacturers enhanced.

Where Modified Atmosphere Packaging has long been obvious for beef and pork, it is now also being used more and more in the poultry industry. With this type of packaging, the air inside the pack is replaced by a protective gas mixture. Apart from extended shelf-life, Modified Atmosphere Packaging also reduces micro bacterial risks for fresh poultry. In addition, the hermetically sealed packs with their accurately cut top film deliver a striking presentation with excellent branding possibilities by means of printing or labelling.

Securely held Grill Specialties

Poultry traditionally has a strong market position during the BBQ season, when marinated specialties are in strong demand. For these kinds of products, modern skin packaging is recommended, which can be achieved with both tray-sealing (TraySkin®) and thermoforming (ThermoSkin®) technology. The poultry product is sealed tightly in the tray or pack by means of a highly transparent barrier skin film, which fits the contours of the product like a second skin. Hence, the product - including its marinade - is securely held in the pack, unable to move and with appetising looks - even in standing or hanging presentation.

Shrink Technology for Optimal Presentation

Whole chickens, due to their complicated shape, have always been a challenge to the industry, but can now be packed in a secure, hygienic and appealing manner. With the ShrinkStyle system, SEALPAC was able to optimise the common shrink packaging process. It starts with the forming of the bottom film, after which the product is loaded.

By loading from above, the common problem of contaminated sealing edges is eliminated. The pack is then evacuated, sealed and dipped in hot water, after which the film fits the contours of the product like a second skin. An optional opening aid ensures consumer-friendly handling of the pack. The extremely strong seal protects the poultry product against drip-loss during the complete production chain, maintains its freshness and clearly extends its shelf-life.

Safe Handling - Meat and Poultry

Safety tips for purchasing, storing, preparing, and cooking meat and poultry. Bacteria are all around us and they love the food we eat just as much as we do. Taking a few simple steps between the store and the dinner plate can prevent a wide range of food borne illnesses. Following are some tips and best practices for handling meat and poultry.

Purchasing and Storing

When purchasing meat in the grocery store, be sure to visit the meat department last. This will minimize the amount of time that the meat is unrefrigerated between purchasing and arriving home. Bacteria grow and multiply rapidly at temperatures between 40 and 140 degrees Fahrenheit (the temperature "danger zone"), so the amount of time spent between these temperatures should be minimized.

Keep meat and poultry separate from other items in your grocery cart. Packaging is not always sealed tight and bacteria filled juices may leak during transport. Preventing meat and poultry packaging from touching other items in your grocery cart, especially fresh produce, reduces cross contamination and the likelihood of food borne illness. Refrigerate meat and poultry immediately upon arriving home. Once again, minimising the amount of time that food spends in the temperature danger zone greatly reduces the risk of bacterial proliferation.

Thawing

Meat and poultry should always be thawed at refrigerated temperatures (40 degrees or lower) and never at room temperature. Thawing in the refrigerator may take longer, but it ensures that all parts of the food remain out of the temperature danger zone during thawing.

Thawing meat and poultry at room temperature may allow the surface, where most of the bacteria resides, to warm up quickly while the inside remains frozen or at safe temperatures. Bacteria can then multiply rapidly on the surface as the meat thaws and cause infection. Meat and poultry can also be thawed under cold running water. Cold running water will keep the exterior of the meat within the temperature safety zone while the interior thaws.

The microwave can also be used to thaw meat and poultry, but this method should only be used if the meat is cooked immediately after thawing. Microwaves tend to heat unevenly and can produce warm spots perfect for bacterial growth. As long as the meat or poultry is cooked immediately after microwaving, the bacteria will not have time to proliferate on these warm spots.

Washing

Meat and poultry should not be washed prior to cooking.

Washing increases the likelihood that bacteria will be splashed onto the surface of your sink, countertop, or other cooking surfaces and cause cross contamination. Any bacteria on the surface of meat or poultry will be killed during the cooking process (if cooked properly) and does not need to be washed off. Washing food is most important for those items that will be consumed raw.

Marinating

Marinating meat and poultry should always be conducted at refrigerated temperatures. Allowing meat and poultry to marinate at room temperature provides bacteria the perfect conditions for growth. Even marinades that are highly acidic allow the growth and proliferation of specific types of bacteria. Marinade used or exposed to raw meat should never be applied to meat after cooking. If needed, separate a portion of the marinade prior to adding the raw meat to keep it bacteria free and safe for use after cooking.

Cook

Meat should always be cooked to the proper temperature to kill harmful bacteria. Ground meat, such as hamburger, should be cooked to the proper temperature both inside and out. Whole cuts of meat should be cooked to the proper temperature on the surface or other areas that may have been exposed to bacteria. Generally, if not cut open, bacteria does not have access to the interior of solid cuts of meat.

Chill

After cooking, meat and poultry should be refrigerated as soon as possible. The less time spent in the temperature danger zone of 40 to 140 degrees, the less chance of infection. Meat and poultry should not be left to cool at room temperature prior to refrigeration. Refrigerators are made to compensate for the increase in temperature when warm items are placed in them.

Although most bacteria are killed during the cooking process, they are quickly introduced again from the environment after cooking. Cooked food left to stand at room temperature provides the perfect breeding ground for bacteria. Cooked food held at room temperature are more at risk for bacterial proliferation than their raw counterparts. Placing items in the refrigerator as soon as possible after cooking is one of the most important steps in food safety.

The Small Flock for Poultry Meat

A well-planned and well-managed flock can be a good source of fresh poultry meat. Large scale commercial broiler production and merchandising techniques often result in market prices difficult to match with a backyard flock. Small flock owners should not plan to produce more birds than the family can use or market, either live or dressed, to friends and neighbours. The most economical meat production is obtained from the commercial meat strains developed from breeds such as the Cornish, Plymouth Rock, and New Hampshire.

These crosses have been bred for the most economical conversion of feed to poultry meat; they feather rapidly and mature early. Some breeds such as White or Barred Plymouth Rocks, Rhode Island Reds, and New Hampshires are used for farm flock meat production, but they generally don't grow as rapidly as the crosses and take more feed per pound of weight gained.

Leghorn males do not make good meat birds and are unprofitable even if day-old chicks are sometimes obtained free. The various classes of chicken meat birds are raised from the same commercial strains. Broilers or fryers are slaughtered at 7-9 weeks of age when they weigh 3-5 pounds and dress a 2½ to 4 pound carcass.

The same bird slaughtered at 5 weeks of age provides a Cornish game hen and can be grown out to 12 weeks or longer to make a delicious roaster. The males can be caponised at 3-4 weeks and marketed after 18 weeks as capons. Meat-type chicks are usually purchased on a straight-run (males and females mixed) basis.

Economic Considerations

Most small flocks produce only enough meat for home consumption. If any birds are to be marketed, consider the prevailing market prices for the various classes of poultry meat birds and don't try to compete with retail sales at distressed market or special sale prices. Many rural customers prefer fryer-type chickens of the heavier weights rather than the lighter weight ranges common in many stores.

Calculate your production costs and compare to retail market prices. Figure your chick cost with a small allowance (a few extra chicks) for mortality. It will take about 5 pounds of feed to age 6 weeks and 8-9 pounds to 8 weeks for the commercial strains. Roasters and capons require more feed per pound of meat produced than fryers. Since feed efficiency is much poorer as the birds get heavier, for the most economical production don't keep birds beyond the desired weights.

Figure any equipment costs depreciated over a 10-year period and housing costs over a 20-year period if these expenses are incurred. Estimate your expenses for litter, heat for brooding, lights, and miscellany. Allow for any payments made for labour for caring for birds, cleaning out house, etc. Convert your figures to a per pound basis by dividing the total cost per bird by the expected market weight. The ready-to-cook weight will be 70-75 percent of the live bird weight.

Housing and Equipment

Fig. Suggested Brooding area Arrangement.

Housing requirements for raising poultry meat birds can be quite minimal from late spring through fall. More substantial housing and more adequate heating equipment will be needed for winter flocks. Almost any small building that meets the floor space requirements for the size of flock desired can be used. A small number of chicks can even be brooded in a corner of a garage. Space requirements are even less if brooding and growing battery units are used. Most poultry meat birds are raised under total confinement, although some roaster and capon flocks are allowed access to limited amount of fenced range or yard area.

Brooding, feeding, and watering equipment can be purchased from local feed and farm supply outfits or from mail-order houses. Much can be home-built equipment. Sizes of feed hoppers should be changed as birds grow so they can easily eat without wasting feed. Hanging tube-type feeders, adjustable in height, can be used as the birds grow. It is desirable to place a platform under water fountains to avoid wet litter. Automatic waterers save labour even with small flocks.

The house and equipment should be thoroughly cleaned and disinfected before starting chicks. If chickens have been in the house previously, remove all litter and wash the house and equipment with water under pressure. Scrub and scrape all organic matter from building and equipment surfaces. After cleaning, disinfect building and equipment using an approved compound and following manufacturer's directions. Dry and air the building, then place 2-4 inches of wood shavings, straw or other litter material on the floor. Place a cardboard fence around the brooding area to confine the chicks to the heat source for the first week.

Brooding Methods

Electric brooders are available for brooding small numbers of chicks. Most of the larger brooder units are fired with either gas or oil. Variations of a simple light bulb brooder can be made from the plan shown in poultry.

When using a brooder, start the chicks at 90-95° F measured 2 inches off the floor under the edge of the hover. Reduce the temperature 5 degrees per week until supplemental heat is no longer needed. Watch the chicks as a guide to their comfort. If the chicks crowd together under the brooder, increase the heat. Lower the temperature if they tend to move away from the heat source. Allow 7 to 10 square inches of space under the brooder for each chick. Provide ½ square foot of brooder house space per chick. Start the brooder the day before the chicks arrive and adjust to proper operating temperature.

Infrared lamps provide a convenient heat source for rearing chicks. Use porcelain sockets approved for these lamps and suspend the lamps with a chain or wire (not the electric cord) so the lamp gets no closer than 15 inches to the litter. If the average brooder house temperature is 50° F, one 250-watt infrared lamp is generally sufficient for heating 80 chicks.

One chick can be added to this estimate for every degree over 50° F; one chick should be subtracted for every degree below 50° F. The use of more than one lamp is recommended so chicks will not be without heat if a lamp burns out. Supply more heat by lowering the lamps to not less than 15 inches above the litter or use more or higher wattage lamps. To reduce heat turn off some lamps, use smaller lamps or raise the lamps to 24 inches above the litter. You are heating the chicks only and not the air so air temperature measurements can't be used as a guide to chick comfort when using infrared lamps.

Place feed on chick box lids or trays from cut down cardboard boxes for the first few days. Feed and water should be available to the chicks as soon as they arrive. Chicks need 1 lineal inch of feeding space for the first 2 weeks, 2 lineal inches for 2-6 weeks, and 3 lineal inches after 6 weeks. Feed wastage is minimized by filling hoppers half full and adjusting feeder height or size to birds' size.

Provide a 1 gallon water fountain for 50 chicks the first 2 weeks. Increase the number of size of waterers from 2-10 weeks to provide 40 inches of watering space per 100 birds or 1 gallon capacity for 15 chicks. Provide 1 gallon capacity per 10 birds for older birds if using fountains or 1 inch of trough space.

Management Recommendations

Floor space per bird should be increased to 1 square foot from 6-10 weeks. From 10 weeks on they will need additional space-at least 2 to 3 square feet if they do not have access to a yard or range. Butchering some of the birds at various ages increases floor space for remaining birds. Maintain litter in good condition, removing wet spots and caked litter as necessary to keep the floor dry and birds comfortable. Some producers reuse litter for 2-3 broods of broilers if no disease or parasite problems have been encountered.

For the small poultry flock a complete feed is most convenient. Farms that have mixing facilities for other live stock operations can use local grains mixed with the appropriate concentrate following the directions of the supplier. A 22-24 percent protein starter mash is usually fed to poultry meat birds for the first 4 weeks.

Many feeding programmes then switch to a 20 percent protein finisher feed until broiler market time. Meat birds grown on chick starter and developer feeds with lower protein and energy content will not gain as rapidly as those on a broiler feeding programme. A small amount of grit can be fed once or twice a week.

Roasters and capons fed on high energy broiler feeding programmes are more prone to leg problems and breast blister development. Birds to be raised to heavier weights can be fed a developer feed with lower protein and energy content following the broiler starter until 2 to 3 weeks before processing. This developer minimizes early fat deposition and provides more adequate skeletal and muscle development during the growing period. A high energy finisher feed is then fed for the final 2 to 3 weeks.

Breast blisters result from an irritation of the tissue in the keel bone area. Maintaining good litter condition, preventing overcrowding, using equipment without sharp edges, and feeding programmes that develop good body structure before heavy weights are reached are all factors that can help reduce incidence of breast blisters.

Cannibalism and feather picking are other problems that may develop with poultry meat birds. Debeaking at the hatchery will eliminate these problems. Various factors such as crowding, nutrient deficiencies, inadequate ventilation, too little drinking and eating space, too much light, idleness, and the appearance of blood on an injured bird contribute to picking. Good management can frequently ward off cannibalism. In small flocks a pick-paste remedy can be used with much success if the problem is not out of hand. Isolation from other birds is a first means of preventing disease. Restrict unnecessary traffic of people and pets to the poultry flock. If different ages of chickens are present, physically separate the flocks as much as possible and care for the younger birds first. It is easier to control diseases and parasites if birds are kept confined.

Obtain chicks that are from pullorum-typhoid clean stock. Coccidiosis can usually be prevented by good sanitation and a low level coccidiostat drug in the feed. Rotate range areas so that birds are not on the same ground year after year. Adjust ventilation to avoid moisture and ammonia build-up in the house. Clean waterers daily and periodically wash them with a sanitising solution. A local veterinarian, county extension agent, or commercial field serviceman can assist with flock health and other management problems or direct you to competent help.

References

Gisbert Otterstatter: *Coloring of Food, Drugs and Cosmetics*, Taylor and Francis Publications, 2010.

H.A. Modi: *Fermented Milk Products*, Aavishkar Publication, Delhi, 2011.

Harmeet Singh: *Dairy Farming*, APH Publication, Delhi, 2011.

7

The Expanded Roles of Food and Beverage Packaging

Introduction

Food and beverage packaging comprises 55% to 65% of the $130 billion value of packaging in the United States. Food processing and packaging industries spend an estimated 15% of the total variable costs on packaging materials. Industrial processing of food, reduced consumption of animal protein, importation of raw materials and ingredients to be converted in the United States, and scarcity of time to select/prepare food from fresh ingredients have enhanced innovation in food and beverage packaging.

The continued quest for innovation in food and beverage packaging is mostly driven by consumer needs and demands influenced by changing global trends, such as increased life expectancy, fewer organizations investing in food production and distribution, and regionally abundant and diverse food supply. The use of food packaging is a socioeconomic indicator of increased spending ability of the population or the gross domestic product as well as regional food availability.

This Scientific Status Summary provides an overview of the latest innovations in food packaging. It begins with a brief history of food and beverage packaging, covering the more prominent packaging developments from the past, and proceeds to more modern advances in the packaging industry.

The article then delves into current and emerging innovations in active and intelligent packaging (such as oxygen scavengers and moisture control agents), packaging mechanisms that control volatile flavours and aromas (such as flavour and odour absorbers), and cutting-edge advances in food packaging distribution (such as radio frequency identification and electronic product codes). Finally, the nano-sized components that have the potential to transform the food packaging industry.

History of Food and Beverage Packaging

Modern food packaging is believed to have begun in the 19th century with the invention of canning by Nicholas Appert. After the inauguration of food microbiology by Louis Pasteur and colleagues in the 19th century, Samuel C. Prescott and William L. Underwood worked to establish the fundamental principles of bacteriology as applied to canning processes. These endeavors to preserve and package food were paralleled by several other packaging-related inventions such as cutting dies for paperboard cartons by Robert Gair and mechanical production of glass bottles by Michael Owens.

In the beginning of the 20th century, 3-piece tin-plated steel cans, glass bottles, and wooden crates were used for food and beverage distribution. Some food packaging innovations stemmed from unexpected sources. For example, Jacques E. Brandenberger's failed attempts at transparent tablecloths resulted in the invention of cellophane. In addition, wax and related petroleum-based materials used to protect ammunition during World War II became packaging materials for dry cereals and biscuits.

Many packaging innovations occurred during the period between World War I and World War II; these include aluminum foil, electrically powered packaging machinery, plastics such as polyethylene and polyvinylidene chloride, aseptic packaging, metal beer cans, flexographic printing, and flexible packaging. Most of these developments helped immeasurably in World War II by protecting military goods and foods from extreme conditions in war zones.

Tin-plated soldered side-seam steel progressed to welded sideseam tin-free steel for cans, and 2-piece aluminum with easy open pop tops were invented for beverage cans, spearheading the exponential growth of canned carbonated beverages and beer during the 1960s and 1970s. The development of polypropylene, polyester, and ethylene vinyl alcohol polymers led the incredible move away from metal, glass, and paperboard packaging to plastic and flexible packaging. Later 20th century innovations include active packaging (oxygen controllers, antimicrobials, respiration mediators, and odour/aroma controllers) and intelligent or smart packaging.

Distribution packaging is already influenced by the potential role of radio frequency identification for tracking purposes. Products such as retort pouches and trays, stand-up flexible pouches, zipper closures on flexible pouches, coextrusion for films and bottles, and an inexorable drive by injection stretch blowmolded polyester bottles and jars for carbonated beverages and water have emerged as rigid and semi-rigid packaging.

Multilayer barrier plastic cans, microwave susceptors, dispensing closures, gas barrier bags for prime cuts of meat, modified atmosphere packaging, rotogravure printed full-panel shrink film labels, and dual oven able trays are examples of innovations for the convenience attributes that have propelled food and beverage packaging into the 21st century.

Moreover, some 21st century innovations are related to nanotechnology whose future may lie in improving barrier and structural/mechanical properties of packaging materials and development of sensing technologies. The principal drivers for most of these innovations have been consumer and food service needs and demands for global and fast transport of food. These packaging innovations are derived largely from industry research and development programmes.

The principal function of packaging is protection and preservation from external contamination. This function involves retardation of deterioration, extension of shelf life, and maintenance of quality and safety of packaged food. Packaging protects food from environmental influences such as heat, light, the presence or absence of moisture, oxygen, pressure, enzymes, spurious odours, microorganisms, insects, dirt and dust particles, gaseous emissions, and so on. All of these cause deterioration of foods and beverages.

Prolonging shelf life involves retardation of enzymatic, microbial, and biochemical reactions through various strategies such as temperature control; moisture control; addition of chemicals such as salt, sugar, carbon dioxide, or natural acids;

removal of oxygen; or a combination of these with effective packaging. Precise integration of the product, process, package, and distribution is critical to avoid recontamination. The ideal packaging material should be inert and resistant to hazards and should not allow molecular transfer from or to packaging materials.

Other major functions of packaging include containment, convenience, marketing, and communication. Containment involves ensuring that a product is not intentionally spilled or dispersed. The communication function serves as the link between consumer and food processor. It contains mandatory information such as weight, source, ingredients, and now, nutritional value and cautions for use required by law. Product promotion or marketing by companies is achieved through the packages at the point of purchase.

Secondary functions of increasing importance include traceability, tamper indication, and portion control. New tracking systems enable tracking of packages though the food supply chain from source to disposal. Packages are imprinted with a universal product code to facilitate checkout and distribution control. More recent innovations used include surface variations sensed by finger tips and palms, sound/music or verbal messages, and aromas emitted as part of an active packaging spectrum. Gloss, matte, holograms, diffraction patterns, and flashing lights are also used.

Active and Intelligent Food Packaging

Traditional food packages are passive barriers designed to delay the adverse effects of the environment on the food product. Active packaging, however, allows packages to interact with food and the environment and play a dynamic role in food preservation. Developments in active packaging have led to advances in many areas, including delayed oxidation and controlled respiration rate, microbial growth, and moisture migration.

Other active packaging technologies include carbon dioxide absorbers/emitters, odour absorbers, ethylene removers, and aroma emitters. While purge and moisture control and oxygen removal have been prominent in active packaging, purge control is the most successful commercially. An example is the drip-absorbing pad used in the poultry industry. In addition, active packaging technology can manipulate permselectivity, which is the selective permeation of package materials to various gases.

Through coating, microperforation, lamination, coextrusion, or polymer blending, permselectivity can be manipulated to modify the atmospheric concentration of gaseous compounds inside a package, relative to the oxidation or respiration kinetics of foods. Certain nanocomposite materials can also serve as active packaging by actively preventing oxygen, carbon dioxide, and moisture from reaching food. Intelligent or smart packaging is designed to monitor and communicate information about food quality. Examples include time-temperature indicators (TTIs), ripeness indicators, biosensors, and radio frequency identification. These smart devices may be incorporated in package materials or attached to the inside or outside of a package.

As of summer 2008, the commercial application of these technologies to food packaging has been small. However, the U.S. Food and Drug Administration (FDA) recognizes TTIs in the 3rd edition of the *Fish and Fisheries Products Hazards and Control Guidance,* so their importance may increase in the seafood industry. Moreover, Wal-Mart, Home Depot and other retail outlets use radio frequency identification, so it is likely to become very prominent as a mechanism for tracking and tracing produce and others perishable commodities.

Oxygen Scavengers

The presence of oxygen in a package can trigger or accelerate oxidative reactions that result in food deterioration: Oxygen facilitates the growth of aerobic microbes and molds. Oxidative reactions result in adverse qualities such as off-odours, off-flavours, undesirable colour changes, and reduced nutritional quality. Oxygen scavengers remove oxygen (residual and/or entering), thereby retarding oxidative reactions, and they come in various forms: sachets in headspace, labels, or direct incorporation into package material and/or closures.

Oxygen scavenging compounds are mostly agents that react with oxygen to reduce its concentration. Ferrous oxide is the most commonly used scavenger. Others include ascorbic acid, sulfites, catechol, some nylons, photosensitive dyes, unsaturated hydrocarbons, ligands, and enzymes such as glucose oxidase. To prevent scavengers from acting prematurely, specialized mechanisms can trigger the scavenging reaction. For example, photosensitive dyes irradiated with ultraviolet light activate oxygen removal. Oxygen scavenging technologies have been successfully used in the meat industry.

Carbon Dioxide absorbers and emitters

Carbon dioxide may be added for beneficial effects, for example, to suppress microbial growth in certain products such as fresh meat, poultry, cheese, and baked goods. Carbon dioxide is also used to reduce the respiration rate of fresh produce and to overcome package collapse or partial vacuum caused by oxygen scavengers.

Carbon dioxide is available in various forms, such as moisture-activated bicarbonate chemicals in sachets and absorbent pads. Conversely, high levels of carbon dioxide resulting from food deterioration or oxidative reactions could cause adverse quality effects in food products. Excess carbon dioxide can be removed by using highly permeable plastics whose permeability increases with higher temperatures.

Moisture control agents

For moisture-sensitive foods, excess moisture in packages can have detrimental results: for example, caking in powdered products, softening of crispy products such as crackers, and moistening of hygroscopic products such as sweets and candy. Conversely, too much moisture loss from food may result in product desiccation.

Moisture control agents help control water activity, thus reducing microbial growth; remove melting water from frozen products and blood or fluids from meat products; prevent condensation from fresh produce; and keep the rate of lipid oxidation in check. Desiccants such as silica gels, natural clays and calcium oxide are used with dry foods while internal humidity controllers are used for high moisture foods (for example, meat, poultry, fruits, and vegetables).

Desiccants usually take the form of internal porous sachets or perforated water-vapour barrier plastic cartridges containing desiccants. They can also be incorporated in packaging material.

Humidity controllers help maintain optimum in-package relative humidity (about 85% for cut fruits and vegetables), reduce moisture loss, and retard excess moisture in headspace and interstices where microorganisms can grow. Purge absorbers remove liquid squeezed or leaking from fresh products and can be enhanced by other active additives such as oxygen scavengers, antimicrobials, pH reducers, and carbon dioxide generators.

Antimicrobials

Antimicrobials in food packaging are used to enhance quality and safety by reducing surface contamination of processed food; they are not a substitute for good sanitation practices. Antimicrobials reduce the growth rate and maximum population of microorganisms (spoilage and pathogenic) by extending the lag phase of microbes or inactivating them.

Antimicrobial agents may be incorporated directly into packaging materials for slow release to the food surface or may be used in vapour form.

Research is underway on the antimicrobial properties of the following agents:

- *Silver Ions*: Silver salts function on direct contact, but they migrate slowly and react preferentially with organics. Research on the use of silver nanoparticles as antimicrobials in food packaging is ongoing, but at least 1 product has already emerged: FresherLongerTM storage containers allegedly contain silver nanoparticles infused into polypropylene base material for inhibition of growth of microorganisms.
- *Ethyl Alcohol*: Ethyl alcohol adsorbed on silica or zeolite is emitted by evaporation and is somewhat effective but leaves a secondary odour.
- *Chlorine Dioxide*: Chlorine dioxide is a gas that permeates through the packaged product. It is broadly effective against microorganisms but has adverse

secondary effects such as darkening meat colour and bleaching green vegetables.

- *Nisin*: Nisin has been found to be most effective against lactic acid and Gram-positive bacteria. It acts by incorporating itself in the cytoplasmic membrane of target cells and works best in acidic conditions.
- *Organic Acids*: Organic acids such as acetic, benzoic, lactic, tartaric, and propionic are used as preservative agents.
- *Allyl Isothiocyanate*: Allyl isothiocyanate, an active component in wasabi, mustard, and horseradish, is an effective broad spectrum antimicrobial and antimycotic. However, it has strong adverse secondary odour effects in food.
- *Spice-based Essential Oils*: Spice-based essential oils have been studied for antimicrobial effects: for example, oregano oil in meat, mustard oil in bread, oregano, basil, clove, carvacol, thymol, and cinnamon.
- *Metal Oxides*: Nanoscale levels of metal oxides such as magnesium oxide and zinc oxide are being explored as antimicrobial materials for use in food packaging.

Ethylene Absorbers and Adsorbers

Ethylene is a natural plant hormone produced by ripening produce. It accelerates produce respiration, resulting in maturity and senescence. Removing ethylene from a package environment helps extend the shelf life of fresh produce. The most common agent of ethylene removal is potassium permanganate, which oxidizes ethylene to acetate and ethanol.

Ethylene may also be removed by physical adsorption on active surfaces such as activated carbon or zeolite. Potassium permanganate is mostly supplied in sachets while other adsorbent or absorbent chemicals may be distributed as sachets or incorporated in the packaging materials.

Temperature Control: Self-heating and Cooling

Self-heating packaging employs calcium or magnesium oxide and water to generate an exothermic reaction. It has been used for plastic coffee cans, military rations, and on-the-go meal platters. The heating device occupies a significant amount of volume (almost half) within the package. Self-cooling packaging involves the evaporation of an external compound that removes heat from contents (usually water that is evaporated and adsorbed on surfaces).

Advances in Controlling Volatile Flavours and Aromas

The mass transfer of components between and within food and packaging leads to the loss of volatile flavours and aromas from food. The most common methods of mass transfer food packaging systems are migration, flavour scalping, selective permeation, and ingredient transfer between heterogeneous parts of the food.

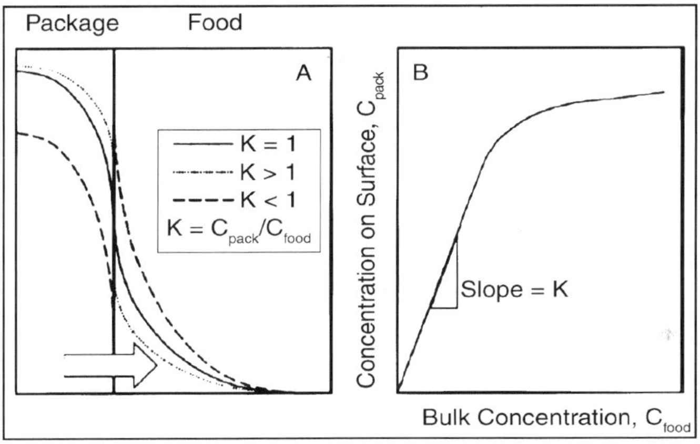

Fig. Effect of Interfacial Partitioning of Molecules on Mass Transfer between Food and Packaging Materials

Notes:

(a) Migration of Package Constituents into Foods,

(b) Scalping of Food Flavour into Packaging Materials. K is a Partition Coefficient of a Migrant

Migration is the transfer of substances from the package into the food due to direct contact. Migration of packaging components to food must be understood and considered with toxicological risk analysis. Most incidences of migration occur in plastic packaging systems; thus, the most commonly studied migrants are plastic monomers, dimers, oligomers, antioxidants, plasticizers, and dye/adhesive solvent residues. Migration of packaging material components is examined in 2 ways, based on the migrating chemicals.

One is global migration (that is, total migration); the other is specific migration of chemicals of interest. Migration of chemical substances is determined by the units of mg/kg for food or mg/m^2 for package surface.

The degree of migration depends on several variables: contact area between food and package material, contact time, food composition, concentration of migrant, storage temperature, polymer morphology, and polarity of polymeric packaging materials and migrants. Flavour scalping is caused by the absorption of desirable volatile food flavours by package materials (for example, absorption of volatile flavours of orange juice and citrus beverages by polyethylene).

Polyethylene materials are known to scalp many volatiles from food. This is due to polyethylene's lipophilic nature, which attracts large amounts of nonpolar compounds such as volatile flavours and aroma in foods. In fact, certain products—especially high-fat foods or vacuum-packaged foods—pick up odours from adjacent strong odour foods when stored or distributed in the same case, storage room, or trailer. The absorption of undesirable flavours by packaging materials follows the same theory and principles of migration but is generally not considered flavour scalping. Unacceptable odour pick-up can be avoided by proper package wrapping with high-barrier materials.

The use of high-barrier packaging materials can also prevent the absorption of other nonfood odours such as taints. Because they result in deterioration of quality and consumer preference of the packaged food, both migration and flavour scalping are unfavourable. However, some applications intentionally utilize these methods of mass transfer to improve the quality of packaged foods. The interactions can be used in active and intelligent packaging applications.

Examples are off-flavour absorbing systems and beneficial volatile release systems. Besides loss of flavour in food, nonpolar flavour components loosen the polymer structure to create more amorphous polymers. This causes undesired changes in the mechanical (seal strength, loss of laminations) and barrier (to oxygen, moisture, and volatiles) properties. Therefore, the effect of off-flavour absorption on essential characteristics of plastic packaging materials should be investigated.

Flavour and Odour Absorbers

In packaged foods, flavour and odour absorbers take in unwanted gaseous molecules such as volatile package ingredients, chemical metabolites of foods, microbial metabolites, respiration products, or off-flavours in raw foods. Examples are sulfurous compounds and amines produced biochemically from protein degradation, aldehydes and ketones produced from lipid oxidation or anaerobic glycolysis, and bitter taste compounds in grapefruit juice. Flavour-and odour-absorbing systems use the same mass transfer mechanism as flavour scalping to remove off-characteristics and are typically available as films, sachets, tapes/labels, and trays.

Flavour and odour absorbers are usually placed inside packages or combined with other flavour permeable materials. Although scavenging of malodourous constituents is recommended to improve the quality of packaged foods, this technology should not be used to mask off-odours produced by hazardous microorganisms that could place consumers at risk. Absorption is the dominant mass transfer phenomenon when porous materials absorb flavour molecules.

Porous or fineparticulate materials have extremely large surface areas; therefore, even with low absorption rates, the overall absorption is very significant. Porous materials—such as zeolites, clays, molecular sieves, active carbon, maltodextrin, and cyclodextrin—have been found to increase flavour. Besides the absorption rate, the maximum absorption, which is also a significant factor, can be controlled by the thickness of films, with thicker films absorbing more flavours. The film thickness does not alter absorption kinetics but relates to the total amount of absorption.

The overall absorption rate that includes the surface adsorption rate and in-structure diffusion rate is affected by molecular size of flavours, polarity of flavour molecules and packaging materials, porosity of packaging materials, storage temperature, morphology of plastic packaging materials (that is, glass transition temperature, crystallinity, and free volume), and relative humidity. Polar plastic materials or corona treatment of film surface and/or high-barrier (or high-density and more-crystalline) materials can be used to prevent nonpolar flavour absorption. When nonporous plastic materials absorb flavour molecules without the occurrence of chemical reactions, the increase in adsorption kinetics and diffusion kinetics will accelerate the absorption rate.

However, when the flavour absorption results in chemical reactions with the plastic material, reaction kinetics as well as adsorption and diffusion kinetics significantly affect the off-flavour elimination capacity. Alkali chemicals such as ammonia and amines can be absorbed and neutralized by packaging materials containing acidic ingredients. This acid–alkali reaction mechanism between volatile chemicals and polymer ingredients has been utilized to eliminate various alkali chemicals and aldehydes.

High Chemical Barrier Material Innovations

High-barrier packaging can significantly reduce adsorption, desorption, and diffusion of gases and liquids to maintain the quality of food. It also prevents the penetration of other molecules such as oxygen, pressurized liquid or gas, and water vapour, which are generally undesirable for food preservation. There are various procedures to enhance the barrier property of packaging materials or packages.

Barrier properties can be improved by combining the package materials with other high-barrier materials through polymer blending, coating, lamination, or metallization. The morphology of the blend relates to its permeability.

Laminar structure (such as coating or lamination) of high-barrier materials on packaging material decreases the permeability linearly with respect to the square thickness; also, blending with platelets or droplets of high-barrier materials reduces permeability but is less effective than coating or lamination at the same mass as that of highbarrier materials. Commercial examples are polyvinylidene chloride (PVdC) coating on oriented polypropylene (OPP), polyethylene terephthalate (PET) lamination on coextruded polypropylene/polyethylene, and aluminum-metallization on PET.

Table. Oxygen Transmission Rate (OTR in cm^3 m^{-2} d^{-1} atm^{-1} at 23 °C, 50% RH) and Water Vapour Transmission Rate (WVTR in g m^{-2} d^{-1} at 23 °C, 75% RH) of Composite Films based on 12 μm PET Films

Film	OTR	WVTR	Specification (μm)
PET	110	15	12
PET/PE	0.93 to 1.24	0.248 to 0.372	12/50
PET/PVDC/PE	0.33	0.132	12/4/50
PET/PVAL/PE	0.13	0.26 to 0.39	12/3/50
PET/EVOH/PE	0.06	0.134 to 0.268	12/5/50
PET/Al-met/PE	0.06 to 0.12	0.006 to 0.03	12/-/50
PET/SiO_x	0.006 to 0.06	0.0024 to 0.06	12/-
PET/Al-foil/PE	0	0	12/9/50

Notes:

PE = polyethylene low density; PVDC = poly(vinylidene chloride); PVAL = poly(vinyl alcohol); EVOH = ethylene vinyl alcohol; Al-met = aluminum metallization; SiO_x = silicon oxide; Al-foil = aluminum foil.

Table shows barrier properties of commercially available laminated (or coated) PET films. Orientation of crystalline polymers such as bi-axially oriented polypropylene (BOPP) and biaxially oriented PET (BOPET) produces lamella structure of polymers.

Other examples are blends of polymers with planar clay particles, a lamella blend structure, and a mixture of beeswax in edible polymer as particulate system films. Other innovative technologies for barrier property enhancement with commercial feasibility include transparent vacuum-deposited or plasma-deposited coating of silica oxide on PET films, epoxy spray on PET bottles, and composites of plastics with nanoparticles.

Factors Affecting Food Packaging Interactions and Barrier Properties

Regardless of the direction of mass transfer (for example, whether flavour absorption or release) or the intent of the mass transfer (for example, whether to achieve desirable transfer or to prevent an undesirable transfer), various factors of food, packaging, and distribution affect the mass transfer kinetics and amount. The transfer of chemicals is principally derived by the concentration difference in food and packaging materials.

A small concentration gradient results in a little transfer while a large gradient results in transfer of a large amount of the compound at a fast rate.

The nature of food is also an important factor: for example, food ingredients like lipids and flavours act as solvents of plastic materials, making them soft, even barrier plastic materials. Lipid and moisture content generally control the transfer of chemicals significantly. Storage temperature, storage duration, and contact surface are also important factors. A high storage temperature accelerates transfer while longer storage duration raises the amount of transfer.

In addition, nonvolatile migrants and flavours can be transferred when the food and packaging are in contact. In such cases, migration time is related to the cumulative duration of contact of migrants to food. Moreover, a package's entire surface area may affect the migration of volatiles and flavours. The use of barrier packaging or inert materials always reduces this type of transfer.

Therefore, simple lamination or coating on packaging material surfaces is a very practical way to modify the barrier properties and, consequently, to control the chemical interactions between food and the package. Use of inert packaging materials with hermetic seals can also be used to contain volatile flavours.

New Advances and Key Areas of Change

Sustainable Food Packaging

Three concepts populate the key areas of change within food packaging. The first is the trend towards more sustainable packaging. While there are multiple definitions of sustainable packaging, the Sustainable Packaging Coalition, an international consortium of more than 200 industry members, offers the most accepted definition.

Sustainable packaging is characterized by the following criteria:

- It is beneficial, safe, and healthy for individuals and communities throughout its life cycle.
- It meets market criteria for performance and cost.

- It is sourced, manufactured, transported, and recycled using renewable energy.
- It maximizes the use of renewable or recycled source materials.
- It is manufactured using clean production technologies and best practices.
- It is made from materials healthy in all probable end-of-life scenarios.
- It is designed to optimize materials and energy.
- It is recovered effectively and used in biological and/or industrial cradle-to-cradle cycles.

Sustainability initiatives led by global legislation, retailers, and corporations guide package material choices, design, and food packaging sales for food packaging professionals. The revised 1997 European Commission's Packaging and Packaging Waste Directive, the 2007 REACH (Registration, Evaluation, and Authorization of Chemicals), and the BS EN 13432 standard (which defines compost-ability, degradability, and biodegradability) are examples of effective global legislative guides in a majority of the Group of 8 countries (Canada, France, Germany, Italy, Japan, Russia, United Kingdom, United States, and the European Union), BRIC (Brazil, Russia, India, and China), and the developing world.

In retail, IKEA and United Kingdom retailers have been long-term promoters of sustainable packaging and have launched impressive initiatives such as Plan A byMarks and Spencer, which defines food packaging material use. Also, Wal-Mart, the world's largest retailer, entered the sustainable packaging arena in 2006. Introduced in 2007, the Wal-Mart scorecard ranks packages compared with category competitors, based on environmental scores.

Finally, because a packaging material's source has been shown to be a defining factor in the final package's sustainability, corporations and environmental coalitions are working together to reduce the effect of packaging materials on global sources. International paper companies, the World Wildlife Fund, the Sustainable Forestry Initiative, and the Forest Stewardship Council verify the sources of wood for use in packaging. Groups such as the Confederation of European Paper Industries (CEPI) in the European Union are addressing solutions to the remaining high environmental impact of paper, which is heavily dependent on water, gas, and energy. Other material-based working groups are taking similar actions.

Use of Packaging Supplier Relationships for Competitive Advantage

The second key area of change in food packaging is the use of packaging-development value chain relationships for competitive advantage. Suppliers to the food packaging industry are adding value in relationships within the traditional package development value chain. This chain extends from raw material generation, to conversion, to production and distribution, to retail, to consumer use and disposal.

But the formerly linear nature of this chain has become an integrated sphere that enables cross-fertilization of ideas fromvarious supply-chain functions directly to the food manufacturer. Investing in packaging supply chain relationships offers opportunities for focus, innovation, and technology transfer for a competitive advantage. For example, Starbucks engaged their packaging value chain—Mississippi River Corp., MeadWestvaco, and the Solo Cup Co.—to create an FDA-approved 10% recycled-fibre coffee cup.

In addition, Green Mountain Coffee Roasters used its supply chain effectively when it introduced a compost able cup (the Ecotainer) with members of their supplier chain: DaniMer Scientific, NatureWorks, and International Paper. Also, Naturipe has used the value chain optimization to enable consistent packaging to stores from over 50 global locations.

Evolution of food service packaging

The third and final key area of change in food packaging is the evolution of food service packaging. Food service has grown to become a major part of consumer spending. As this trend increases, packaging plays a key role in ensuring food safety and providing convenience to consumers. For instance, proper package labeling allows food preparers to know the source of a food, its proper holding temperature, and the adequate cooking needed.

Ease of package opening allows for fewer utensils needed to open packages, which lessens contamination. While tracking and shelf-life extension technologies are employed in the food service industry to reduce the risk of food borne illness, proper heating and heat retention continue to be challenges. Consequently, packaging's role in ensuring proper heating and heat retention continues to increase. For example, CuliDish is a new product that uses varying levels of aluminum within a tray to package foods that require heating together with those that do not require heating.

The tray allows foods requiring high heat to be heated in the microwave at the same time with foods that do not require heat (such as salads). The food can be served in the same tray, thus reducing handling. More innovations in the areas of heat and heat retention are expected to assist in reducing food safety risks associated with improper cooking.

Two major convenience trends—meals eaten in transit and multi-component meals—have also advanced the food service packaging industry. The popularity of meals eaten in transit is evident since only 60% of meals are prepared and eaten at home and 20% of consumers eat on their way to another location instead of in a fixed position. On-the-go food consumption has resulted in packaging that contains a greater variety of foods.

Technologies that support this trend include edible films to enrobe food particles and edible wraps that peel off, allowing consumers to eat numerous foods while in transit. Food package design innovations such as modular folding cartons with flip-off lids, pouches that are easily opened or have a seatbelt flap to hold food close to consumers, and reusable packaging have contributed to the increase of eating while on the go. The growth of multi-component meals in food service stems from multiple types of foods being ordered at quick-service restaurants. This innovation allows for ease of food service preparation and less waste.

Multi-component packaging also provides consumers with a presentation platform that makes multi-component foods easy to consume. Examples include reusable trays that clip together into 1 larger serving tray such as that used in the Les Petits Grande line of products; KFC's triple dip strip cartons, which allow consumers to select from 1 of 3 dips when dipping chicken strips; and the folding tray that allows consumers to carry 2 cups of coffee with 1 hand.

Advances in Food Packaging Distribution

RFID Systems for Packaged Foods: Architecture and Working Principles

Radio frequency identification (RFID) is a system that uses radio waves to track items wirelessly. RFID makes use of tags or transponders (data carriers), readers (receivers), and computer systems (software, hardware, networking, and database). The tags consist of an integrated circuit, a tag antenna, and a battery if the tag is passive (most active tags do not require battery power).

The integrated circuit contains a non-volatile memory microchip for data storage, an AC/DC converter, encode/decode modulators, a logic control, and antenna connectors. The wireless data transfer between a transponder/tag and a reader makes RFID technology far more flexible than other contact identifications, such as the barcode system and thus makes it ideal for food packaging.

The working principles of an RFID system are as follows:

- Data stored in tags are activated by readers when the objects with embedded tags enter the electromagnetic zone of a reader;
- Data are transmitted to a reader for decoding; and
- Decoded data are transferred to a computer system for further processing.

Tag frequency is related to the working principles of an RFID system (for example, magnetic coupling or electric coupling) and the reading range. Frequency depends on the type of tag, reader, and cost. The typical RFID frequencies are low frequency, high frequency, ultra high frequency, and microwave frequency. Generally, low frequency systems have short reading ranges, slow read speeds, and lower cost while higher frequency RFID systems are utilized when longer read ranges and fast reading speeds are required. Microwave frequency requires active RFID tags.

Electronic Product Codes

In 2004, Electronic Product Code (EPC) Global Network began developing a second generation RFID protocol: EPC Class 1 version 2. The main goal of Gen 2 is to create a single global standard that is compatible with ISO standards. The tags would work in various countries that have different commercial band frequency. EPC is the most significant function of RFID contributing to commercial industry.

EPC improves the traceability of items and facilitates efficient product recall and authenticity. EPC is similar to Universal Product Code (UPC), which is commonly used in bar codes. Compared to UPC, which uses 12 digits of numbers, EPC has 64 to 256 bits of alphanumeric data. The most common EPC has 96 bits. The first obstacle for wide utilization of RFID is the cost for tags. Tags are still too expensive for use on individual primary packages.

The infrastructure required for RFID systems (including readers, database servers with communication systems, and other information technology to process huge amounts of data) is costly and needs to be shared with all users in supply chains. The global use of EPC also requires compatibility among various regulations and standards of radio frequency.

The biggest hurdle to wide utilization of EPC for RFID is the potential problems of privacy protection. A hidden reading system could collect all data from tags of items and also RFID card holders for the purpose of data stealing or data removing. Guidelines for the ethical use of RFID systems for data collecting, data handling, and system security need to be established.

RFID for the Food Industry

RFID has recently found its way into numerous applications in the food industry, ranging from food monitoring and traceability to enhancing food safety, to improving supply chain efficiency. The major benefits of RFID technology in the food industry are greater speed and efficiency in stock rotation and better tracking of products throughout the chain, resulting in improved on-shelf availability at the retail level and enhanced forecasting. The technology is well suited for many operations in food manufacturing and supply chain management. An RFID-based resource management system can help users handle warehouse operating orders by retrieving and analysing warehouse data, which could save time and cost. The use of RFID in the food industry is currently focused on tracking and identification. When RFID technology becomes more established in the food industry, the integration of food science knowledge will be necessary to develop the intelligent food packaging application for food quality and safety.

Some food companies have already integrated RFID into manufacturing and distribution. Retail chains such as Wal-Mart and Home Depot have been testing the technology for distribution. In 2003, Wal-Mart issued a mandate requiring its top 100 suppliers to use RFID tags on all cases and pallets entering its distribution centers by 2005.

RFID compliance is a long-term project for Wal-Mart; more of its suppliers are expected to be compliant by the end of 2008. Other major players advocating RFID technology are the U.S. Department of Defence and major retailers such as Albertsons, Target, Tesco, and Marks and Spencer. RFID technology also provides security and safety benefits for food companies through tracking the origin of supplies. For example, a small California winery uses RFID to track its barrels and to enhance wine making by streamlining data collection.

The company planted RFID tags on tanks and harvesting bins, allowing better control of wine production and tracking. In addition, by attaching an RFID tag to a package, the package becomes intelligent because the stored data provide valuable information that can be stored and read by appliances. This intelligent packaging technology is also being extended to refrigeration and freezing. Appliances can communicate with the packages and identify information related to the storage of the packaged products. Despite these benefits, other factors such as cost of the technology and recycling ability need to be considered.

Supply Chain Management, Traceability, and Recall

As a tool for tracking, RFID is a promising technology to food supply chain management. The FDA, 1307 recalls of processed foods occurred between 1999 and 2003; these recalls could have been avoided with a technology such as RFID. Exposure to risk at any one of the stages of the processed food supply chain would result in a domino-effect breakdown that could affect the smooth running of an entire supply chain.

All stages of the supply chain (from farming, processing, transportation, manufacturing, retailing, and warehousing to consumption) are equally critical to food recall problems. If RFID technology were combined with Hazard Analysis and Critical Control Point systems, the supply chain stages would be integrated, traceable, and effectively managed by food processors to reduce the number of recalls significantly. The outstanding tracing abilities of RFID tags to individual food product could enable manufacturers to audit every single phase of a product in a retail unit, monitoring correct handling, transportation, storage, and delivery.

New Tag for Use on Metal and Metal Packaging and High Water Content Products

The RFID system is not infallible; it has some weaknesses, such as the shielding effect of metal, which affects signal transduction. Data cannot be read correctly when tags are attached to metal on the surface or inside the package. Until recently, high frequency RFID tags were used on metal beer barrels. However, low frequency RFID (125 kHz to 135 kHz) has less loss of radio signal by the metal materials in the high magnetic field of a reader compared to higher frequency signals.

Another issue is that water molecules can absorb microwave signals, resulting in signal loss or interference during data acquisition from microwave RFID tags. Since most foods contain high moisture, this signal interference requires further study to enable application of the technology in the food industry. It is worth noting that low, high, and ultra-high frequency (UHF) tags can be used in high water systems since water does not interfere with their signal; on the contrary, ice absorbs UHF radio signals.

For example, a product such as ice cream (which is technically defined as partly frozen foam with ice crystals and air bubbles occupying a majority of the space) contains about 72% frozen water and thus interferes with UHF radio signals. Ice cream manufacturers have overcome this product interference by placing tags over an air gap in the containers. Companies with diverse product lines and high levels of automation will likely encounter significant technical barriers to RFID implementation.

Nanocomposites and Other Emerging Nanotechnologies in Food Packaging

Nanotechnology has the potential to transformfood packaging materials in the future. Such nanoscale innovation could potentially introduce many amazing new improvements to food packaging in the forms of barrier and mechanical properties, detection of pathogens, and smart and active packaging with food safety and quality benefits.

The nanolayer of aluminum that coats the interior of many snack food packages is one common example of the role that nanotechnology already plays in food packaging. The market for nanotechnology in food packaging in 2006 was estimated at $66 million and is expected to reach $360 million in 2008.

Nanomaterials are abundant in nature and numerous techniques are available to fabricate various nanomaterials. Nanoparticles can be produced top down from larger structures by grinding, use of lasers, and vaporization followed by cooling. Alternately, bottom-up methods are commonly used for synthesis of complex nanoparticles. These methods include solvent extraction/evaporation, crystallization, self-assembly, layer-by-layer deposition, microbial synthesis, and biomass reactions. All of these are being researched for potential application in food packages in the future. One group of nanomaterials at the forefront of food packaging development is nanocomposites.

Nanocomposites

Nanocomposite packages are predicted to make up a significant portion of the food packaging market in the near future. Principia Markets, a consulting firmthat tracks the plastics market, estimates that the market for nanocomposites will reach 1 billion pounds by 2010. Many nanocomposite food packages are either already in the marketplace or being developed. The majority of these are targeted for beverage packaging. In large part, the impetus for this predicted growth is the extraordinary benefits nanoscience offers to improve food packages. Improvements in fundamental characteristics of food packaging materials such as strength, barrier properties, antimicrobial properties, and stability to heat and cold are being achieved using nanocomposite materials.

In the late 1980s, Toyota was the 1st company to commercialize nanocomposite materials. They found that the addition of 5%-by-weight nano-sized montmorillonite clay significantly increased the mechanical and thermal properties of different grades of nylon. Nanocomposite materials are now used in gasoline tanks, bumpers, and interior and exterior panels. Research on use of nanocomposites for food packaging began in the 1990s.

Most of the research has involved the use of montmorillonite clay as the nanocomponent in a wide range of polymers such as polyethylene, nylon, polyvinyl chloride, and starch. Amounts of nanoclays incorporated vary from 1% to 5% by weight. Nanocomponents must have 1 dimension less than 1 nm wide. The lateral dimensions, on the other hand, can be as large as several micrometers, leading to high aspect ratios (ratio of length to thickness) of many of these materials.

The high surface area results in unique properties when nanocomposites are incorporated into packages. There are 3 common methods used to process nanocomposites: solution method, *in situ* or interlamellar polymerization technique, and melt processing.

The solution method can be used to form both intercalated and exfoliated nanocomposite materials. In the solution method, the nanocomposite clay is first swollen in a solvent. Next, it is added to a polymer solution, and polymer molecules are allowed to extend between the layers of filler.

The solvent is then allowed to evaporate. The *in situ* or interlamellar method swells the fillers by absorption of a liquid monomer. After the monomer has penetrated in between the layers of silicates, polymerization is initiated by heat, radiation, or incorporation of an initiator. The melt method is the most commonly used method due to the lack of solvents. In melt processing, the nanocomposite filler is incorporated into a molten polymer and then formed into the final material.

Generally, there are 3 possible arrangements for layered silicate clay nanocomposite materials:

1. Nonintercalated,
2. Intercalated, and
3. Exfoliated or delaminated.

In nonintercalated materials the polymer does not fit between the layered clay, leading to a microphase separated final structure. In intercalated systems, the polymer is located between clay layers, increasing interlayer spacing. Some degree of order is retained in parallel clay layers, which are separated by alternating polymer layers with a repeated distance every few nanometers. Exfoliated systems achieve complete separation of clay platelets in random arrangements. This is the ideal nanocomposite arrangement but is hard to achieve.

Bayer produces transparent nanocomposite plastic films and coatings called Durethan, which contains clay nanoparticles dispersed throughout the plastic. Large amounts of silicate nanoparticles are interspersed in polyamide films. These nanoparticles block oxygen, carbon dioxide, and moisture from reaching fresh meats and others foods. The nanoclay particles act as impermeable obstacles in the path of the diffusion process, thereby extending the shelf life of foods while improving their quality.

The final package is also considerably lighter, stronger, and more heat-resistant. In years past, packaging beer in plastic bottles was not possible due to oxidation and flavour problems. Recently, however, this challenge has been overcome using nanotechnology. For example, Nanocor, a subsidiary of Amcol International Corp., is producing nanocomposites for use in plastic beer bottles that facilitate a 6-mo shelf life. By combining the nanocomposite and oxygen scavenger technologies, a new family of barrier nylons was recently developed for use in multilayer, co-injection blow-molded PET bottles.

In the near future, nanocrystals embedded in plastic bottles may increase beer shelf life up to 18 mo by minimizing loss of carbon dioxide from and entrance of oxygen into bottles. Similar materials are being developed to extend the shelf life of soft drinks. Another advantage of these nanocomposite bottles is that theirweight is considerably less, thereby reducing transportation costs. A considerable amount of research is also occurring in the area of biodegradable nanocomposite food packages.

By pumping carbohydrates and clay fillers through high shear cells, films can be produced with exfoliated clay layers. These films act as very effective moisture barriers by increasing the tortuosity of the path water must take to penetrate the films. Significant increases in film strength are also frequently achieved in these types of materials. Starch and chitosan are two of the most studied biodegradable matrices. In the future, these types of biodegradable nanocomposite food packages may be found in the marketplace.

Other Nanotechnologies

Carbon nanotubes are cylinders with nanoscale diameters that can be used in food packaging to improve its mechanical properties. In addition, it was recently discovered that they may also exert powerful antimicrobial effects. *Escherichia coli* died immediately upon direct contact with aggregates of carbon nanotubes. Presumably, the long, thin nanotubes punctured the *E. coli* cells, causing cellular damage. Single-walled carbon nanotubes may eventually serve as building blocks for antimicrobial materials. Nano-wheels were also recently developed to improve food packaging.

Inorganic alumina platelets have been self-assembled into wagon-wheel shaped structures that are incorporated into plastics to improve their barrier and mechanical properties. This was the first time large wheel-shaped molecules had been formed. The addition of nanosensors to food packages is also anticipated in the future. Nanosensors could be used to detect chemicals, pathogens, and toxins in foods.

Numerous research reports describe detection methods for bacteria, viruses, toxins, and allergens using nanotechnology. For example, adhering antibodies to *Staphylococcus* enterotoxin B onto poly(dimethyl-siloxane) chips formed biosensors that have a detection limit of 0.5 ng/mL. Nanovesicles have been developed to simultaneously detect *E. coli* 0157:H7, *Salmonella* spp., and *Listeria monocytogenes*. Liposome nanovesicles have been devised to detect peanut allergen proteins. In addition, AgroMicron has developed a NanoBioluminescence detection spray containing a luminescent protein that has been engineered to bind to the surface of microbes such as *Salmonella* and *E. coli*.

When bound, it emits a visible glow that varies in intensity according to the amount of bacterial contamination. This product is being marketed under the name BioMark. Nanosensors Inc. is another company pursuing this potential. Through a license agreement with Michigan State University, a nanoporous silicon-based biosensor has been developed to detect *Salmonella* and *E. coli*.

A prototype nanobiosensor was recently tested to detect *Bacillus cereus* and *E. coli* and was found to be able to detect multiple pathogens faster and more accurately than current devices. Finally, Mahadevan Iyer and his colleagues at Georgia Institute of Technology are experimenting with integrating nanocomponents in ultra-thin polymer substrates for RFID chips containing biosensors that can detect foodborne pathogens or sense the temperature or moisture of a product. DNA biochips are already under development to detect pathogens.

Researchers at the Univ. of Pennsylvania and Monell Chemical Sciences Center have used nano-sized carbon tubes coated with strands of DNA to create nanosensors with abilities to detect odours and tastes. A single strand of DNA serves as the sensor and a carbon nanotube functions as the transmitter. Using similar technologies, electronic tongue nanosensors are being developed to detect substances in parts per trillion, which could be used to trigger colour changes in food packages to alert consumers when food is spoiled.

A unique aspect of these biochips is that the DNA is self-assembled onto the chips and repairs itself if damaged. In addition, researchers at Cornell Univ. have invented synthetic DNA barcodes to tag pathogens and monitor pathogens. The nanobarcodes fluoresce under ultraviolet light when target compounds are detected. Another colour-changing film that could find its way into food packages is polymer opal films. Scientists at the United Kingdom's Univ. of Southampton and the Deutsches Kunststoff Inst. in Germany developed these unique self-assembled structures from arrays of spheres stacked in 3 dimensions. Polymer opal films belong to a class of materials known as photonic crystals. The crystals are built of tiny repeating units of carbon nanoparticles wedged between spheres, leading to intense colors that mimic the colors associated with the photonic crystals found on butterfly wings and peacock feathers. Photonic crystals could be used to produce unique food packaging materials that change colour.

Conclusion

The food industry has seen great advances in the packaging sector since its inception in the 18th century with most active and intelligent innovations occurring during the past century. These advances have led to improved food quality and safety. While some innovations have stemmed from unexpected sources, most have been driven by changing consumer preferences. The new advances have mostly focused on delaying oxidation and controlling moisture migration, microbial growth, respiration rates, and volatile flavours and aromas.

This focus parallels that of food packaging distribution, which has driven change in the key areas of sustainable packaging, use of the packaging value chain relationships for competitive advantage, and the evolving role of food service packaging. Nanotechnology has potential to influence the packaging sector greatly. Nanoscale innovations in the forms of pathogen detection, active packaging, and barrier formation are poised to elevate food packaging to new heights.

References

Hema Thapar: *Nutrition and Food Science,* Pacific Books International, Delhi, 2011.

J. O. Dahl, and J. H. Breland: *Food Standards Handbook for Quantity Cookery,* Whitlock Press, Stamford Conn, 1945.

J. O. Dahl: *200 Ways to Control Food Costs in Quantity Cookery,* The Dahls, Stamford Conn, 1941.

8

Dairy Process Biotechnology and Products Packaging

Introduction

Food processing makes use of various unit operations and technologies to convert relatively bulky, perishable and typically inedible raw materials into more useful shelf-stable and palatable foods or potable beverages. Processing contributes to food security by minimizing waste and losses in the food chain and by increasing food availability and marketability. Food is also processed in order to improve its quality and safety. Food safety is a scientific discipline that provides assurance that food will not cause harm to the consumer when it is prepared and/or eaten according to its intended use.

Biotechnology as applied to food processing in most developing countries makes use of microbial inoculants to enhance properties such as the taste, aroma, shelf-life, texture and nutritional value of foods. The process whereby micro-organisms and their enzymes bring about these desirable changes in food materials is known as fermentation.

Fermentation processing is also widely applied in the production of microbial cultures, enzymes, flavours, fragrances, food additives and a range of other high value-added products. These high value products are increasingly produced in more technologically advanced developing countries for use in their food and non-food processing applications. Many of these high value products are also imported by developing countries for use in their food-processing applications.

This document will discuss the prospects and potential of applying biotechnology in food processing operations and to address safety issues in food systems with the objective of addressing food security and responding to changing consumer trends in developing countries. It is important to note that food safety evaluation or risk assessment will not be discussed here. Instead, this chapter will focus on the context of biotechnologies as applied to food safety.

Technologies applied in the processing of food must assure the quality and safety of the final product. Safe food is food in which physical, chemical or microbiological hazards are present at a level that does not present a public health risk. Safe food can, therefore, be consumed with the assurance that there are no serious health implications for the consumer. Recent food scares such as mad cow disease and the melamine contamination of food products have increased consumer concern for food safety. As incomes rise, consumers are increasingly willing to pay a premium for quality, safety and convenience.

A range of technologies is applied at different levels and scales of operation in food processing across the developing world. Conventional or "low-input" food processing technologies include drying, fermentation, salting, and various forms of cooking, including roasting, frying, smoking, steaming, and oven baking. Low-income economies are likely to employ these as predominant technologies for the processing of staple foods.

Many of these technologies make use of a simple, often rudimentary, technological base. Medium levels of processing technologies such as canning, oven drying, spray drying, freeze drying, freezing, pasteurization, vacuum packing, osmotic dehydration and sugar crystallization are widely applied in middle-and upper middle-income economies.

Higher-level, more capital-intensive food-processing technologies such as high-temperature short-time pasteurization and high-pressure low-temperature food processing are widely employed in middle-and upper middle-income economies. Functional additives and ingredients produced using fermentation processes are generally incorporated into food-processing operations that make use of higher-level technologies. Traditional methods of food-safety monitoring such as the detection of pathogenic bacteria in food are generally based on the use of culture media. These are the techniques of choice in low-and lower-middle-income economies which lack the resources, infrastructure and technical capacity to utilize modern biotechnological techniques.

Conventional bacterial detection methods are time-consuming multi-step procedures. At least two to three days are required for the initial isolation of an organism, followed by the requirement for several days of additional confirmatory testing. Biotechnology-based methods can provide accurate results within a relatively short time frame.

Biotechnological developments have resulted in the widespread availability of low-cost rapid methods of identification when compared with the significant cost/time requirements of conventional techniques. Lower-middle-income economies apply both traditional and more sophisticated methods for monitoring the microbiological quality of foods and their conformance to international standards.

A number of case studies are described in the text to demonstrate the utility of biotechnology-based applications in food processing and food safety. These case studies provide the basis for the development of strategic interventions designed to upgrade food processing and food safety in developing countries through the application of biotechnology.

Stocktaking: Learning from the Past

Biotechnology: Definition and scope

For the purpose of this chapter, biotechnology is defined in accordance with the Convention on Biological Diversity, *i.e.* "*any technological application that uses biological systems, living organisms, or derivatives thereof, to make or modify products or processes for specific use*".

Biotechnology in the food processing sector makes use of micro-organisms for the preservation of food and for the production of a range of value-added products such as enzymes, flavour compounds, vitamins, microbial cultures and food ingredients. Biotechnology applications in the food-processing sector, therefore, target the selection and manipulation of micro-organisms with the objective of improving process control, product quality, safety, consistency and yield, while increasing process efficiency. Biotechnological processes applicable to the improvement of microbial cultures for use in food-processing applications include traditional methods of genetic improvement such as classical mutagenesis and conjugation.

These methods generally focus on improving the quality of micro-organisms and the yields of metabolites. Hybridization is also used for the improvement of yeasts involved in baking, brewing and in beverage production. *Saccharomyces cerevisiae* strains have, for example, been researched for improved fermentation, processing and biopreservation abilities, and for capacities to increase the wholesomeness and sensory quality of wine.

Recombinant gene technology, the best-known modern biotechnology, is widely employed in research and development for strain improvement. The availability of genetic manipulation tools and the opportunities that exist to improve the microbial cultures associated with food fermentations are tempered by concerns over regulatory issues and consumer perceptions. *Genetically modified* (GM) microbial cultures are, however, used in the production of enzymes and various food-processing ingredients such as monosodium glutamate, polyunsaturated fatty acids and amino acids.

Biotechnology is also widely employed as a tool in diagnostics in order to monitor food safety, prevent and diagnose food-borne illnesses and verify the origins of foods. Techniques applied in the assurance of food safety focus on the detection and monitoring of hazards whether biological, chemical or physical. These applications will be explored and discussed in subsequent sections.

Current Status of the Application of Traditional and New Biotechnologies in Food Processing in Developing Countries

Methods of Microbial Inoculation in Food Fermentations

The fermentation bioprocess is the major biotechnological application in food processing. It is often one step in a sequence of food-processing operations, which may include cleaning, size reduction, soaking and cooking. Fermentation bioprocessing makes use of microbial inoculants for enhancing properties such as the taste, aroma, shelf-life, safety, texture and nutritional value of foods. Microbes associated with the raw food material and the processing environment serve as inoculants in spontaneous fermentations, while inoculants containing high concentrations of live micro-organisms, referred to as starter cultures, are used to initiate and accelerate the rate of fermentation processes in non-spontaneous or controlled fermentation processes. Microbial starter cultures vary widely in quality and purity.

Starter culture development and improvement is the subject of much research both in developed and in developing countries. While considerable work on GM starter culture development is ongoing at the laboratory level in developed countries, relatively few GM micro-organisms have been permitted in the food and beverage industry globally. In 1990, the United Kingdom became the first country to permit the use of a live genetically modified organism (GMO) in food. It was a baker's yeast, engineered to improve the rate at which bread dough rises by increasing the efficiency with which maltose is broken down.

This modification was done by using genes from yeast and placing them under a strong constitutive promoter. The United Kingdom has also approved a GM brewer's yeast for beer production. By introducing a gene encoding glucoamylase from yeast, better utilization of carbohydrate present in conventional feedstock can be obtained, resulting in increased yields of alcohol and the ability to produce a full-strength, low-carbohydrate beer. More recently, two genetically modified yeast strains were authorized for use in the North American wine industry.

Current literature documents volumes of research reports on the characterization of microbes associated with the production of traditional fermented foods in developing countries. Relatively few of these studies document the application of the diagnostic tools of modern biotechnology in developing and designing starter cultures.

The development and improvement of microbial starters has been a driving force for the transformation of traditional food fermentations in developing countries from an "art" to a science. Microbial starter culture development has also been a driving force for innovation in the design of equipment suited to the hygienic processing of traditional fermented foods under controlled conditions in many developing countries.

Starter culture improvement, together with the improvement and development of bioreactor technology for the control of fermentation processes in developed countries, has played a pivotal role in the production of high-value products such as enzymes, microbial cultures, and functional food ingredients. These products are increasingly produced in more advanced developing economies, and are increasingly imported by less advanced developing countries, as inputs for their food processing operations.

Spontaneous Inoculation of Fermentation Processes

In many developing countries, fermented foods are produced primarily at the household and village level, using spontaneous methods of inoculation. Spontaneous fermentations are largely uncontrolled. A natural selection process, however, evolves in many of these processes which eventually results in the predominance of a particular type or group of micro-organisms in the fermentation medium.

A majority of African food-fermentation processes make use of spontaneous inoculation. Major limitations of spontaneous fermentation processes include their inefficiency, low yields of product and variable product quality. While spontaneous fermentations generally enhance the safety of foods owing to a reduction of pH, and through detoxification, in some cases there are safety concerns relating to the bacterial pathogens associated with the raw material or unhygienic practices during processing.

"Appropriate" Starter Cultures as Inoculants of Fermentation Processes

"Appropriate" starter cultures are widely applied as inoculants across the fermented food sector, from the household to industrial level in low-income and lower-middle-income economies. These starter cultures are generally produced using a backslopping process which makes use of samples of a previous batch of a fermented product as inoculants.

Appropriate starter cultures are widely applied in the production of fermented fish sauces and fermented vegetables in Asia and in cereal or grain fermentations in African and Latin American countries. The inoculation belt used in traditional fermentations in West Africa serves as a carrier of undefined fermenting micro-organisms, and is one example of an appropriate starter culture.

It generally consists of a woven fibre or mat or a piece of wood or woven sponge, saturated with "high"-quality product of a previously fermented batch. It is immersed into a new batch, in order to serve as an inoculant. The inoculation belt is used in the production of the indigenous fermented porridges, "uji" and "mawe", as well as in the production of the Ghanaian beer, "pito".

Iku, also referred to as iru, is yet another example of an "appropriate" starter culture produced by backslopping. This starter culture is produced from concentrated fermented dawadawa, mixed with ground unfermented legumes, vegetables such as pepper, and cereals, such as ground maize. It is stored in a dried form and is used as an inoculant in dawadawa fermentations in West Africa.

A range of appropriate starter cultures, either in a granular form or in the form of a pressed cake is used across Asian countries as fermentation inoculants. These traditional mould starters are generally referred to by various names such as *marcha* or *murcha* in India, *ragi* in Indonesia, *bubod* in the Philippines, *nuruk* in Korea, *koji* in Japan, *ragi* in Malaysia and *Loog-pang* in Thailand. They generally consist of a mixture of moulds grown under non-sterile conditions.

Defined Starter Cultures as Inoculants of Fermentation Processes

Few defined starter cultures have been developed for use as inoculants in commercial fermentation processes in developing countries. Nevertheless, the past ten years have witnessed the development and application of laboratory-selected and pre-cultured starter cultures in food fermentations in a few developing countries. These developments have taken place primarily in Asian countries. "Defined starter cultures" consist of single or mixed strains of micro-organisms. They may incorporate adjunct culture preparations that serve a food-safety and preservative function. Adjunct cultures do not necessarily produce fermentation acids or modify texture or flavour, but are included in the defined culture owing to their ability to inhibit pathogenic or spoilage organisms.

Their inhibitory activity is due to the production of one or several substances such as hydrogen peroxide, organic acids, diacetyl and bacteriocins. By and large, defined cultures are produced by pure culture maintenance and propagation under aseptic conditions.

They are generally marketed in a liquid or powdered form or else as a pressed cake. Loog-pang, a defined culture marketed in Thailand in the form of a pressed rice cake, consists of *Saccharomyces cerevisiae*, *Aspergillus oryzae* or *Rhizopus* sp. and Mucor. Loog-pang has a shelf life of 2–3 days at ambient temperature and 5–7 days under refrigerated conditions.

Ragi cultures are commercially produced by the Malaysian Agricultural Research and Development Institute by mixing a culture inoculum which generally consists of *Rhizopus Oligosporus* with moistened sterile rice flour, and incubating it at ambient temperature for four days.

This starter has a shelf life of two weeks under refrigerated conditions. It is widely used as an inoculant in the production of traditional Malaysian fermented foods. Ragi-type starter cultures for the production of a range of fermented Indonesian products such as oncom, tape and tempeh are currently marketed via the internet.

Defined starter cultures are also widely imported by developing countries for use in the commercial production of dairy products such as yogurt, kefir, cheeses and alcoholic beverages. Many of these cultures are tailored to produce specific textures and flavours. In response to growing consumer interest in attaining wellness through diet, many yogurt cultures also include probiotic strains.

Probiotics are currently produced in India for use as food additives, dietary supplements and for use in animal feed. Methodologies used in the development and tailoring of these starters are largely proprietary to the suppliers of these starters. Monosodium glutamate and lactic acid, both of which are used as ingredients in the food industry, are produced in less-advanced developing countries using defined starter cultures.

Defined Starter Cultures Developed Using the Diagnostic Tools of Advanced Biotechnologies

The use of DNA-based diagnostic techniques for strain differentiation can allow for the tailoring of starter cultures to yield products with specific flavours and/or textures. *Random amplified polymorphic DNA* (RAPD) techniques have been applied in, for example, Thailand, in the molecular typing of bacterial strains and correlating the findings of these studies to flavour development during the production of the fermented pork sausage, nham.

The results of these analyses led to the development of three different defined starter cultures which are currently used for the commercial production of products having different flavour characteristics.

GM Starter Cultures

To date, no commercial GM micro-organisms that would be consumed as living organisms exist. Products of industrial GM producer organisms are, however, widely used in food processing and no major safety concerns have been raised against them. Rennet which is widely used as a starter in cheese production across the globe is produced using GM bacteria. Thailand currently makes use of GM *Escherichia coli* as an inoculant in lysine production.

Many industrially important enzymes such as α-amylase, gluco-amylase, lipase and pectinase and bio-based fine chemicals, such as lactic acid, amino acids, antibiotics, nucleic acid and polysaccharides, are produced in China using GM starter cultures. Other developing countries which currently produce enzymes using recombinant micro-organisms include Cuba, Brazil, India, and Argentina.

Food Additives and Processing Aids

Enzymes, amino acids, vitamins, organic acids, polyunsaturated fatty acids and certain complex carbohydrates and flavouring agents used in food formulations are currently produced using GM micro-organisms.

Enzymes

Enzymes occur in all living organisms and catalyze biochemical reactions that are necessary to support life. They are commonly used in food processing and in the production of food ingredients. The use of recombinant DNA technology has made it possible to manufacture novel enzymes that are tailored to specific food processing conditions. Alpha amylases with increased heat stability have, for example, been engineered for use in the production of high-fructose corn syrups.

These improvements were accomplished by introducing changes in the α-amylase amino acid sequences through DNA sequence modifications of the α-amylase genes. Bovine chymosin used in cheese manufacture was the first recombinant enzyme approved for used in food by the US Food and Drug Administration. The Phospholipase A1 gene from *Fusarium venenatum* is expressed in GM *Aspergillus oryzae* to produce the phospholipase A1 enzyme used in the dairy industry for cheese manufacture to improve process efficiencies and cheese yields.

Considerable progress has been made in recent times towards the improvement of microbial strains used in the production of enzymes. Microbial host strains developed for enzyme production have been engineered to increase enzyme yields by deleting native genes encoding extracellular proteases. Certain fungal producing strains have also been modified to reduce or eliminate their potential for producing toxic metabolites.

Enzymes used in food processing have historically been considered non-toxic. Some characteristics arising from their chemical nature and source, such as allergenicity, activity-related toxicity, residual microbiological activity and chemical toxicity are, however of concern. These attributes of concern must, however, be addressed in light of the growing complexity and sophistication of the methodologies used in the production of food-grade enzymes.

Safety evaluation of all food enzymes, including those produced by GM micro-organisms, is essential if consumer safety is to be assured. Enzymes produced using GM micro-organisms wherein the enzyme is not part of the final food product have specifically been evaluated by the *Joint FAO/ WHO Expert Committee on Food Additives* (JECFA). Safety evaluations have been conducted using the general specifications and considerations for enzyme preparations used in food processing. Preparations of asparaginase enzymes have also been evaluated by JECFA.

Flavours, Amino Acids and Sweeteners

Volatile organic chemicals such as flavours and aromas are the sensory principles of many consumer products and govern their acceptance and market success. Flavours produced using micro-organisms currently compete with those from traditional agricultural sources. Berger more than 100 commercial aroma chemicals are derived using biotechnology either through the screening for overproducers, the elucidation of metabolic pathways and precursors or through the application of conventional bioengineering.

Recombinant DNA technologies have also enhanced efficiency in the production of non-nutritive sweeteners such as aspartame and thaumatin. Market development has been particularly dynamic for the flavour enhancer glutamate which is produced by the fermentation of sugar sources such as molasses, sucrose or glucose using high-performance strains of *Corynebacterium glutamicum* and *Escherichia coli.* Amino acids produced through biotechnological processes are also of great interest as building blocks for active ingredients used in a variety of industrial processes.

Current Status of the Application of Traditional and New Biotechnologies in Food Safety and in Quality Improvement in Developing Countries

Food Safety Issues and Concerns in Food Fermentation Processing

Microbial activity plays a central role in food fermentation processes, resulting in desirable properties such as improvements in shelf-life and quality attributes such as texture and flavour. Pathogenic organisms are, however, of prime concern in fermented foods. Anti-nutritional factors such as phytates, tannins, protein inhibitors, lectins, saponins, oligosaccharides and cyanogenic glucosides are naturally occurring components of raw materials commonly used in food fermentations in developing countries. Contamination of the fermentation process can pose a major health risk in the final fermented product.

Methodologies for identifying and monitoring the presence of chemical and biochemical hazards in fermented foods are, therefore, a critical need. Furthermore, with growing consumer interest in the credence attributes of the products that they consume, and the premium currently being placed on quality linked to geographical origin, the traceability of foods with selected properties is of increasing importance.

Advances in Microbial Genetics

In recent times, the genetic characterization of micro-organisms has advanced at a rapid pace with exponential growth in the collection of genome sequence information, high-throughput analysis of expressed products *i.e.*, transcripts and proteins and the application of bioinformatics which allows high throughput comparative genomic approaches that provide insights for further functional studies.

Genome sequence information, coupled with the support of highly advanced molecular techniques, have allowed scientists to establish mechanisms of various host-defensive pathogen counter-defensive strategies and have provided industry with tools for developing strategies to design healthy and safe food by optimizing the effect of probiotic bacteria, the design of starter culture bacteria and functional properties for use in food processing.

Characterization of the genomes of lactic acid probiotics has, for example, shed light on the interaction of pathogens with lactic acid bacteria. Nucleotide sequences of the genomes of many important food microbes have recently become available. *Saccharomyces cerevisiae* was the first food microbe for which a complete genome sequence was characterized.

This was followed by genome sequencing of the related yeast, *Kluyveromyces lactis* as well as filamentous fungi which are major enzyme producers and have significant applications in the food-processing industry. Genome nucleotide sequences of many Gram-positive bacteria species have also been completed. The *Bacillus subtilis* genome was the first to be completed, followed by that of the *Lactococcus lactis* genome. Genome sequences of food-borne pathogens such as *Campylobacter jejuni*, verocytotoxigenic *Escherichia coli* O157:H7 and *Staphylococcus aureus* have also been completed. Genome sequences of microbes that are of importance in food processing, such as *Lactobacillus plantarum* are also available. The genome of *Clostridium botulinum,* responsible for food poisoning, was also recently completed.

Detection of Pathogens

The rapid detection of pathogens and other microbial contaminants in food is critical to assess the safety of food products. Traditional methods to detect food-borne bacteria often rely on time-consuming growth in culture media, followed by isolation, biochemical identification, and sometimes serology. Recent technological advances have improved the efficiency, specificity and sensitivity of detecting micro-organisms. Detection technologies employ the *polymerase chain reaction* (PCR) assay.

Short fragments of DNA (probes) or primers are hybridized to a specific sequence or template, which is subsequently enzymatically amplified by the Taq polymerase enzyme using a thermocycler. In theory, a single copy of DNA can be amplified a million-fold in less than 2 hours with the use of PCR techniques; hence, the potential of PCR to eliminate or greatly reduce the need for cultural enrichment.

The genetic characterization of genome sequence information has further facilitated the identification of virulence nucleotide sequences for use as molecular markers in pathogen detection.

Multiplex real-time PCR methods are now available to identify the *E. coli* O157:H7 serogroup. PCR-based identification methods are also available for *Vibrio cholerae* and for major food-related microbes such as *Campylobacter jejuni, C. coli, Yersinia enterocolitica, Hepatitis A virus, Salmonella, Staphylococcus aureus.*

Sophisticated cultural media such as chromogenic or fluorogenic media are not readily used in low-income economies but are relatively widespread in lower-middle-income and upper-middle-income economies. The use of immunoassays such as enzyme-linked immunosorbent assay (ELISA) is also very limited in low-income economies but is more widespread in the form of diagnostic kits in lower-middle and upper-middle-income economies.

DNA methods, which require elaborate infrastructure and high technical competence, find minimal application in lower-income and some lower-middle-income economies. There are movements towards implementing safety-control programmes such as the application of Hazard Analysis and Critical Control Point (HACCP) in food fermentations in many developing countries.

The application of HACCP necessitates the deployment of good agricultural practices, good manufacturing practices (GMPs) and good hygienic practices (GHPs) and the monitoring of critical control points for potential microbial and chemical contamination during bioprocessing. Rigorous adherence to sanitary practices in the processing environment necessitates rapid, dynamic, sensitive, specific as well as versatile and cost-effective assay methods.

The molecular approach of biotechnology entails near-time or real-time bacterial detection, and offers sensitivity and specificity unchallenged by traditional/conventional methods.

Mycotoxin Detection

The problem of mycotoxin contamination in food including fermented foods is a global concern. Mycotoxin contamination is particularly prevalent in developing countries in tropical areas such as in South Asia and Africa.

High-performance liquid chromatography (HPLC) and gas chromatography/mass spectrometry (GC/MS) are two of the most widely used methods for the detection and quantification of mycotoxins in developing countries.

These methods, however, are time-consuming, difficult to use and require laboratory facilities. Immunoassays that are economical in use, sensitive and easy to use would greatly facilitate the detection and quantization of mycotoxins.

A number of ELISA kits are now commercially available for the detection of aflatoxins, deoxynivalenol, fumonisins, ochratoxins, and zearalenone.

Detection and Identification of Foods and Food Ingredients

The DNA-based identification code system is reliant on polymorphisms at the nucleotide level for the differentiation of living organisms at the variety and species levels. Currently PCR-based methods are used either for the purpose of detecting single nucleotide polymorphisms (SNPs) giving rise to restriction fragment length polymorphisms (RFLPs) or for detecting small sequence length polymorphisms (SSLPs) often known as Variable Number Tandem Repeats (VNTRs).

These methods facilitate the identification of unique polymorphisms of a variety of food commodities and can be used in the identification of their source or origin. These unique polymorphisms are often referred to as DNA barcodes. The DNA barcode is used for the identification of specific varieties in food detection and in food traceability.

The DNA barcode has been used for the identification of many products for export in countries such as Thailand, China, Brazil, Cuba and Argentina, The DNA barcode of microsatellite markers has also been successfully used in differentiating and identifying fermented products such as premium wines, cheeses and sausages on the basis of their origins. Basmati rice varieties and olive cultivars used in olive oil production have also been differentiated.

Analysis of the Reasons for Successes/Failures of Application of Biotechnologies in Developing Countries

Socio-economic factors have played a major role in the adoption and application of microbial inoculants in food fermentations. In situations where the cost of food is a major issue, uptake and adoption of improved biotechnologies has been generally slow. Demand for improved inoculants and starter culture development has been triggered by increasing consumer income, education and new market opportunities.

Socio-economics of the Consumer Base

The consumer base of traditionally fermented staple foods in most developing countries is largely poor and disadvantaged. Price, rather than food safety and quality, is therefore a major preoccupation of this group when purchasing food. Fermented foods provide that target group with an affordable source of food, and make a substantial contribution to their food and nutritional security. These foods are generally produced under relatively poor hygienic conditions at the household and village level. Fermentation processing is practised largely as an art in such contexts.

Interventions designed to upgrade processes used in the production of these traditionally fermented staples have been largely carried out through donor-funded projects and have focused primarily on reducing the drudgery associated with the fermentation processes. Improvements have also targeted the upgradation of hygienic conditions of fermentation processes and the introduction of simple and "appropriate" methodologies for the application of inoculants, such as the use of backslopping.

While the uptake of simple backslopping technologies at the household level has, in general, been very good by that target group, the uptake of defined starter cultures has been less successful, owing to cost considerations. Case Study 4.3 on the household level production of Som Fug in Thailand highlights the poor uptake of improved starter culture technologies by household-level processors, primarily on the basis of cost.

With growing incomes and improved levels of education in urban centres across a number of developing countries, dietary habits are changing and a wider variety of foods is being consumed. Fermented foods are no longer the main staples, but are still consumed as side dishes or condiments by that target group.

The demand of that target group for safe food of high quality has begun to re-orient the traditional fermented food sector, and led to improvements in the control of fermentation processes through the development and adoption of defined starter cultures, the implementation of GHPs and HACCP in food fermentation processing, and the development of bioreactor technologies, coupled with appropriate downstream processing to terminate fermentation processes and thus extend the shelf-life of fermented foods. The packaging of fermented products has also improved. Case Study 4.1 on soy sauce production in

Thailand highlights an example of how starter culture development coupled with bioreactor technology has improved yields and the efficiency of fermentation processes, while Case Study 4.2 highlights how consumer demand for safe food led to research and development into starter culture development designed to improve the safety of nham in the marketing chain.

Changing Consumer Demand Trends

Apart from their changing dietary patterns and their demand for safety and quality, higher-income consumers demand convenience and are increasingly concerned about deriving health benefit from the foods they consume. Many of these consumers also show a preference for shopping in supermarkets.

Consumer demand for deriving wellness through food consumption has stimulated the development of industrial fermentation processes for the production of functional ingredients such as polyunsaturated fatty acids and pro-biotic cultures for use as food ingredients in developing countries. These functional ingredients are currently applied in the fortification of fermented foods as well as in the production of dietary supplements in countries such as India.

The growth of supermarkets in developing countries has promulgated the need for standardized products of a reasonable shelf-life that meet safety and quality criteria. Packaged fermented products such as kimchi, miso and tempeh, for example, are widely available in supermarkets across Asia.

The production of traditional beer in a powdered format and in ready-to-drink containers in Zambia is a very good example of product development that has taken place in response to consumer demand for convenience, both in local and export markets.

Shifting consumer preferences in South Africa, away from basic commodity wine to top-quality wine, is yet another example of how market demand has led to research and biotechnological innovation in the wine industry. Biotechnological innovations in that country are currently focused on the improvement of *Saccharomyces cerevisiae* strains to improve wholesomeness and sensory quality of wines.

The Enabling Environment for Starter Culture Development

A considerable amount of research in developing countries has focused on the identification of starter micro-organisms associated with the fermentation of these staple foods. The greatest strides in starter culture development have, however, been realised in countries that have prioritized the development of technical skills, the infrastructural support base and funding support for research into the upgradation of fermentation processes.

Linkages between research institutions and the manufacturing sector have also been critical to the successful introduction of starter cultures. Case Study 4.1 on soy sauce production exemplifies how success was achieved through such collaboration. Case Study 4.2 on nham production in Thailand also highlights how collaboration between the manufacturing sector and public sector research institutions resulted in the development of improved starter cultures and the uptake of these cultures by nham manufacturers to assure product safety.

Collaborative initiatives among research institutions have also had a major positive impact on biotechnological developments in developing countries. Collaboration among African institutions and their counterparts in the North has greatly facilitated improvements in biotechnological research and capacity development in the area of food biotechnology on the continent.

One major success story in this regard has been collaborative projects involving Burkina Faso, Ghana and institutions in the Netherlands. This programme facilitated the typing and screening of microbial cultures associated with fermented African foods as a basis for starter culture development. Results of this work led to improvements in the production of gari, a fermented cassava product and dawadawa, a fermented legume product. Issues related to the protection of intellectual property rights (IPR) are of growing concern with respect to starter culture development.

Proactive Industrial Strategies

Biotechnology developments have been most successful in areas where proactive approaches are taken by industry. The

Thai food industry successfully creates perceived quality by launching new product logos and associating these new products with biotechnology or with the fact that they were developed using traditional biotechnology, such as starter cultures. The goal of the industry is to project an image of itself as producing products of superior quality and safety that represent progressiveness based on a higher level of technology.

Export Opportunities for Fermented Products

Increasing travel due to globalization has changed the eating habits of consumers across the globe. Export markets for fermented foods have grown out of the need to meet the requirements of developing country diaspora in these markets as well as to satisfy growing international demand for niche and ethnic products. Indonesian *tempe* and Oriental soy sauce are well known examples of indigenous fermented foods that have been industrialized and marketed globally.

The need to assure the safety and quality of these products in compliance with requirements of importing markets has been a driving force for the upgrading of starter cultures as well as for diagnostic methodologies for verification of their quality and safety. Growing interest and trade in fermented food products is also likely to lead to the greater use of the DNA barcode for identifying the origins of specific fermented food products produced in developing countries.

Case Studies of Applications of Biotechnologies in Developing Countries

Fermented Soy Sauce Production

This Case Study on the production of soy sauce highlights success in the application of starter culture technology and the use of improved bioreactor technology. It exemplifies the transition of a craft-based production system to a technology-based production system. Research leading to these developments was supported by an international organization, followed by funding support from the Government of Thailand and the Thai soy sauce industry. Developments of the process were largely driven by the demand pull created by a soy sauce industry consortium in Thailand in order to meet market requirements.

Soy sauce production involves a two-step fermentation process that makes use of koji inoculants in the initial phase, followed by moromi inoculants in the second phase. The initial phase of the fermentation involves the soaking of soybeans in water for 1–2 hours, boiling for approximately 17 hours to hydrolyze the protein complex, and the addition of the koji culture *Aspergillus oryzae* for proteolysis of the soy proteins.

Using this traditional method of production, the process of proteolysis takes between 40 hours and seven days depending on the method and the conditions used. The second phase of the fermentation process, which is referred to as a moromi fermentation, involves the addition of brine solution to the koji. *Saccharomyces rouxii,* a salt-tolerant yeast, is the predominant micro-organism in this phase of the fermentation, which lasts as long as 8–12 months.

Moromi fermentations are traditionally conducted in earthenware jars, which often poses a limitation to the manufacturers both in terms of expansion and in terms of production capacity. The soy sauce industry has moved up the ladder of development, from an "art" to a technology-based process through the introduction of defined starter cultures and improvements in the control of the fermentation process. Physical and biological parameters of the fermentation process are controlled through the use of koji and moromi cultures and koji and moromi fermentors.

Use of the koji starter, *Aspergillus flavus* var. *Columnaris,* was found to enhance product safety and uniformity. The introduction of pressure cookers as an innovation for hydrolyzing the soybeans reduced the time required for solubilization from 17 hours to 2.5 hours.

Moreover, the use of starter culture technology facilitated the development of fermentation chambers with controlled temperature and humidity conditions, which resulted in shortening the duration of the fermentation process.

The resulting soy sauce had a higher (6%) soluble protein content than that derived from boiled soybeans. These developments resulted in economic gain for the soy sauce industry and greater value added to the product in terms of quality and safety

Traditional Fermented Pork Sausage (Nham)

Nham is an indigenous fermented pork sausage produced in Southeast Asia. It is prepared from ground pork, pork rinds, garlic, cooked rice, salt, chili, sugar, pepper and sodium nitrite. This Case Study on nham demonstrates how consumer demand for safe food resulted in the commercial use of defined starter cultures, with the collaboration and support of government agencies. The diagnostic role of biotechnology in starter culture development for the tailor making of cultures is also highlighted.

Nham is traditionally consumed as a condiment in the uncooked state in Thailand. It is generally produced using a uncontrolled fermentation process. Fermentation of the product occurs during transportation from the manufacturer to the point of retail. The product is generally retailed under ambient conditions. Traditionally produced nham is considered high risk by Thai health authorities, who require a warning label stating that the product "must be cooked before consumption" on the package.

The first step in the transition to science-based technology for nham fermentations was the development of a starter culture. This starter was subsequently adopted by a nham manufacturer who also implemented HACCP in his operation in order to assure safety and to satisfy the compulsory standard requirements of GMP in the food processing industry imposed by the Thai Food and Drug Administration.

A microbiological hazard profile was developed for nham by the manufacturer in collaboration with scientists from the Ministry of Science, who established that the prevalent pathogens in nham were *Salmonella* spp. (16%), *Staphylococcus aureus* (15%) and *Listeria monocytogenes* (12%). Nitrite, an additive used in nham production was identified as a chemical hazard and the metal clips used for closure of the package were identified as physical hazards. A HACCP plan which included four critical control points was developed for nham.

The critical control point on nitrite was monitored by checking the pre-weighed nitrite prior to adding it to the product formulation. Scientific data generated through the conduct of studies on starter cultures showed that a rapid increase in acidity within 36–48 hours of fermentation inhibited the growth of bacterial pathogens such as *Staphylococcus aureus* and *Salmonella* spp.

With the application of these starter cultures, the final product was sent to retailers after the fermentation reached its end-point (pH< 4. 5). An innovative pH indicator which undergoes a colour change on attainment of the end-point of the fermentation process (pH<4. 5) was included in the package. With these innovations and the implementation of a HACCP plan, local health authorities waived the requirement for the warning "must cook before consumption" on the package.

This authorization was seen by the public as an endorsement of product quality and safety by the health authority. Subsequent to these developments, three medium-sized manufacturers followed suit in adopting the improved technology. Recognition of the starter culture technology as a food safety measure by the health authority was, of itself, an effective public awareness campaign.

RAPD markers were used for the molecular typing of approximately 100 bacterial strains at 12-hour intervals during nham fermentations. These studies resulted in the development and commercialization of three different starter formulae for use by larger manufacturers of nham. These starter cultures are marketed in a liquid form which requires refrigeration.

Dried starter cultures have a shelf life of one month at ambient temperature. Further innovations have led to the incorporation of local yeast extracts into starter culture development, resulting in a 20-to 30-fold reduction in cost. The adoption of starter culture technology in nham fermentations has had a positive impact on the industry in terms of safety assurance to consumers and product consistency.

Traditional Fermented Fish Paste–Som Fug

Som Fug is a traditional fermented minced fish cake. It is considered a healthy and highly nutritious product, and is an excellent source of protein (protein content: 15.7%, fat: 3.2% and total carbohydrates: 4%). It is produced using a spontaneous microbial fermentation process similar to that used for producing nham and many other Southeast Asian fermented foods. This Case Study demonstrates that the uptake and use of starter culture technologies is still largely contingent on cost considerations and consumer appreciation of the nutritional value of the product.

Compositionally, Som Fug consists of minced freshwater fish (mud carp, *Cirrhina microlepis*) 84% (w/w), garlic 8%, water 4%, salt 2%, boiled rice 1%, sucrose 0.1% and black pepper. It is fermented for about 2–4 days at ambient temperature. Lactic acid bacteria are the dominant microflora associated with the fermentation. RAPD-PCR analyses determined that the garlic fermenting lactic acid bacteria associated with Som Fug fermentations belonged to *Lactobacillus pentosus* and *Lact. plantarum*.

Furthermore, the studies also concluded that fructans from garlic are important carbon sources which catalyze the fermentation of Som Fug. The studies of Som Fug illustrate the high discriminatory power of biotechnology in differentiating lactic acid bacteria at the strain level. The Som Fug industry did not see the benefit of implementing starter-culture technology.

Although the important micro-organisms for Som Fug fermentation had been identified, there were no attempts to develop starter cultures. One major reason for the lack of development of starter-culture technology was the widespread production of Som Fug at the household level. Household manufacturers do not see the benefit of starter-culture technology but, rather, view starter cultures as a burden to the cost of production. Moreover, there is no scientific information to substantiate the nutritional value of Som Fug and hence there is very little public awareness of the nutritional value of the product.

Flavour Production from Alkaline-fermented Beans

This Case Study on the indigenous fermentation of the locust bean, dawadawa (fermented locust bean), is a classic example of how traditional fermentations can be exploited for the production of high-value products such as flavour compounds. The work, however, was undertaken by a large cooperation with little involvement of local researchers. Returns on commercial successes derived from this study did not go back to the people who invented the traditional method of producing this indigenous fermented food. This Case Study, therefore, serves to highlight the critical issue of IPR of traditional production systems. Dawadawa is produced by alkaline fermentation of the African fermented locust bean. It is an important condiment in the West/ Central African Savannah region. Similar fermented food products can be found throughout Africa, with regional differences in the raw materials used as processing inputs or in postprocessing operations.

Similarly fermented products are referred to as "kinda" in Sierra Leone, "iru" in coastal Nigeria, "soumbara" in Gambia and Burkina Faso, and "kpalugu" in parts of Ghana. Foods produced by alkaline fermentation in other parts of the world include "natto" in Japan, "*thua noa*" in Thailand and "kinema" in India. These are mainly used as culinary products to enhance or intensify meatiness in soups, sauces and other prepared dishes. The production of dawadawa involves extensive boiling and dehulling of the beans, followed by further boiling to facilitate softening. Spontaneous fermentation of the softened beans is subsequently allowed to take place over 2–4 days. Micro-organisms associated with the fermentation include *Bacillus subtilis*, *B. pumilus*, *B. licheniformis* and *Staphylococcus saprophyticus*.

During the fermentation process, the pH increases from near neutral to approximately 8.0, temperature increases from 25 °C to 45°C and moisture increases from 43% to 56%. At the same time, a five-fold increase in free amino acids takes place, and glutamate, a flavour enhancer, increases five-fold during the process. Mechanisms of flavour production during the fermentation process, as well as flavour principles generated during dawadawa fermentation processing, have been studied by international food manufacturers and been used as a basis for the development of flavours for incorporation in bouillon-type products.

Looking Forward: Preparing for the Future

A Key Issue in the Sector where the Application of Biotechnologies could be useful

The identification of infectious agents requires high-end

technologies which are not usually available in developing countries. Developing countries must, therefore, seek assistance from countries with higher calibre technologies in order to characterize the infectious agents, put in place surveillance and monitoring systems and develop strategies to contain the disease(s). Biotechnology can play a key role in facilitating the characterization of new emerging pathogens.

Traditional cultural methods for the detection and enumeration of microbial pathogens are tedious and require at least 12–18 hours for the realization of results. By that time, the food products would have been distributed to retailers or consumers. Immunoassay diagnostic kits facilitate near-real-time monitoring, sensitivity, versatility and ease of use. The emergence of multi-antibiotic resistance traits is prevalent in intensive farming in developing countries due to the abuse of antibiotics.

The spread of multi-antibiotic resistant micro-organisms poses public health concerns, because pathogens exhibiting such resistance would be difficult to control with the use of currently available antibiotics.

The rapid detection of these pathogens, with high sensitivity, is one way of monitoring and containing the spread of multi-antibiotic resistant traits. A strategic approach being employed by some is the development of affinity biosensors with an antibiotic resistant nucleotide sequence as the detection probe.

Identifying options for Developing Countries

It is important that countries recognize the potential of fermented foods and prioritize actions to assure their safety, quality and availability. Based on the stocktaking exercise in this document, a number of specific options can be identified for developing countries to help them make informed decisions regarding adoption of biotechnologies in food processing and in food safety for the future.

Regulatory and Policy Issues

- Governments must be committed to protecting consumer health and interests, and to ensuring fair practices in the food sector.

- There has to be consensus at the highest levels of government on the importance of food safety, and the provision of adequate resources for this purpose.
- Government policy that is based on an integrated food-chain approach is science-based, transparent and includes the participation of all the stakeholders from farm to table must be put in place.
- The importance of the regional and international dimensions of the use of biotechnologies in food processing and safety must be recognized.
- Priority must be accorded to promoting fermented foods in the food-security agendas of countries.
- Governments must also provide an enabling environment that is supportive of the growth and development of upstream fermentation processes such as the production of high-value fermented products, such as enzymes, functional-food ingredients and food additives.

International Cooperation and Harmonization

- The organization and implementation of regional and international fora are critical requirements for the enhancement of national organizational capability and performance and for the facilitation of international co-operation. Further, the setting up of administrative structures with clearly defined roles, responsibilities and accountabilities could efficiently govern processed foods and safety issues.
- Biotechnology-based Standard Operating Procedures (SOPs) for food safety should also be documented for use in authorized laboratories.
- National food control databases for the systematic collection, reporting and analysis of food-related data (food inspection, analysis, etc.) with set regulations and standards based on sound science and in accordance with international recommendations (Codex) are key requirements.

Education Policy

- While the consumption of fermented foods is growing in popularity among higher-income consumers thanks to increasing interest in wellness through diet, the consumption of fermented foods by lower-income consumers in many developing countries is perceived to be a backward practice.
  - Strategies should therefore be developed for the dissemination of knowledge about food biotechnology and, particularly, fermented foods. Targeted consumer education on the benefits of consuming fermented food products and on applying good practice in their production is required.
  - Food biotechnology should be included in educational curricula in order to improve the knowledge base in countries on the contribution of fermented foods to food and nutritional security and to generate awareness of the growing market opportunities for fermented foods and high-value products derived from fermentation processes.

Information-sharing

Access to specialized technical information on biotechnology and biotechnological developments in the food processing sector are critical and necessary inputs and support systems for guiding and orienting the research agendas of countries. The necessary information systems should therefore be developed to facilitate rapid access to information on biotechnological developments across both the developed and the developing world.

Legislation and Policy on Technologies

Expertise in legislation and technology licensing, as well as knowledge about how to nurture innovation and turn it into business ventures, are critical requirements for developing countries. Successful technology transfer requires all of these elements and an environment that is conducive to innovation. Government policy in developing countries should therefore prioritize technology transfer that helps create new business ventures, an approach that requires government support such as tax incentives and infrastructure investment.

Intellectual Property Rights (IPR)

- Many of the traditional fermentation processes applied in developing countries are based on traditional knowledge. Enhanced technical and scientific information is required in order to claim ownership of the traditional knowledge of the craft of indigenous fermented foods. Lack of technical knowledge has resulted in the failure to realise the benefits of the industrialization of indigenous fermented foods by individuals who are the rightful owners of the technology.
- Greater focus is required on issues of relevance to IPR and on the characterization of microbial strains involved in traditional fermentation processes. Emphasis must be placed on IPR education for scientists. National governments should put in place the requisite infrastructure for IPR to facilitate the process. At the institutional level, this infrastructure would include technology management offices for assisting scientists in procedures relating to intellectual property matters. The processes used in the more advanced areas of agricultural biotechnology are generally covered by IPR, and the rights are generally owned by parties in developed countries.

Communication and Consumer Perceptions

- Communication between various stakeholders is critical in proactively engaging with consumers. Communication must be established with the public at large on processed food and associated hazards. Communication gradually builds confidence and will be critical to advancing the application of biotechnologies in food processing and safety. The primary role of communication in this respect is to ensure that information and opinions from all stakeholders are incorporated in the discussion and decision-making process. The need for specific standards or related texts and the procedures followed to determine them should also be clearly outlined. The process, therefore, should be transparent.

- Public awareness and education are critical to the success of food bioprocessing and food safety in developing countries.
- Greater attention must be directed towards understanding consumer and producer (processor) perceptions on food safety and quality in developing countries.
- If foods are to be promoted as being safe and healthy, their nutritional and safety attributes must be transparently demonstrated by presenting scientific data to substantiate the nutritional and health benefits and by applying good manufactu-ring/hygiene practice and HACCP as safety measures to ensure that issues of consumer concern are addressed.

Technical Capacities and Technology Transfer

- Traditional fermented foods should be viewed as valuable assets. Governments should capitalize on these assets and add value to them by supporting research, education and development, while building on and developing the indigenous knowledge base on food fermentations.
  - Government agencies in developing countries should focus on the development of technical capacities to deal with emerging technical issues.
  - The technical capacities of academic and research institutes should be strengthened in the fields of food biotechnology, food processing, bioprocess engineering and food safety through training and exchange programmes for researchers. Such programmes should emphasize collaboration with both developed and developing country institutions engaged in work on food biotechnology, starter culture development, bioprocess engineering and food safety.

- Training capabilities in food biotechnology and food safety should be developed within developing country institutions through the introduction of degree courses in order to broaden the in-country technical support base for food bioprocess development. Given the similarities among fermentation processes across regions, an inventory of institutions engaged in food biotechnology in developing countries would be an asset in facilitating networking among institutions. Food processors, policy-makers and equipment manufacturers should also be integrated into the networking activities.
- The development of appropriate levels of bioreactor technology with control bioprocess parameters will be necessary to improve the hygienic conditions of the fermentation processes.
- Research and infrastructural development to enable the cost-effective production of defined starter cultures in a stable format (*i.e.* cultures which do not require refrigeration and have prolonged shelf-life under ambient conditions) should be prioritized.
- Infrastructure development to facilitate the transfer and adaptation of fermentation technologies developed at the laboratory level to the household and village and, where necessary, the enterprise level should be prioritized.
- Appropriate levels of equipment will also be required to facilitate the downstream processing of these products.
- Traceability systems that facilitate the differentiation and identification of food products should be prioritized in order to broaden market opportunities for these products.
- A food-chain approach to assuring food safety should be prioritized by governments.

• Food safety management systems should be strengthened by implementing systematic food safety measures such as GHP, GMP and HACCP in food fermentation operations. Diagnostic kits are important tools for monitoring and verifying the level of sanitation in processing plants.

- Highly sensitive and rapid diagnostic kits are invaluable for monitoring and rapidly detecting chemical and microbiological hazards that pose a threat to human health, with high precision and sensitivity. The development of low-cost diagnostic kits suitable for use by small processors would greatly facilitate food-safety monitoring. Development should target the realization of multiplex diagnostic systems with the capacity to detect several pathogens or many chemical contaminants using a single diagnostic kit. The development of diagnostic kits at a national level could further reduce their cost of production. Given the regional specificity of bacterial pathogens at the species and subspecies levels, such diagnostic kits should be developed with specificity and sensitivity to the species or subspecies that are prevalent in a specific region. Investment is therefore needed for the development of expertise, facilities and the infrastructure for the mass production of antibodies, cell culture technology and for the formation of technical know-how on assembling the requisite components of diagnostic kits.
- The development of national hazard-profile databases that document the prevalent pathogens in different regions will be critical. Such information would be useful for further research into the development of diagnostic kits with high precision and sensitivity and in implementing HACCP as well as risk assessment research. The culture collection of identified infectious agents in the hazard profiles could play an important role for specific antibody production for use in the development of immunoassay diagnostic kits.

Identifying Priorities for Action for the International Community

The last decade has witnessed considerable change with respect to the application of biotechnology in food processing and food-safety applications. Market forces have been the major drivers of change in the food sector of developing countries.

Modern biotechnological tools are likely to play a greater role in the development of efficient science-based processes for food processing and safety in order to respond to consumer demand. The production of high-value fermented products such as enzymes, functional food ingredients and food additives is likely to continue to increase in developing countries.

The international community (FAO, UN organizations, NGOs, donors and development agencies) can play a major role in assisting developing countries to maximize the benefit to be derived from food bioprocessing.

The adoption of biotechnology-based methods in food processing and for food safety and quality monitoring is dependent on several factors that include capacity-building in technical and regulatory areas, policy formulation, regulatory frameworks and regional networks. On the basis of analysis in this document, a number of priority areas to be supported by the international community are presented below.

Capacity-building and Human Resource Development

- Support for basic and advanced education.
- Prioritization of specific areas for investment.
- Development of policies, priorities and action programmes that promote food fermentation as a means of addressing food security.
- Support for human resource development in a range of scientific disciplines–food biotechnology, food safety, bioengineering and enzyme technology.
- Support for capacity-building initiatives for household-level, small-and edium-scale processors of fermented foods.
- Support for IPR development.

Technology Transfer and Support for Research and Development

- Improvement of the relevance of national research to the needs of the food sector in developing countries.
- Enhancement of competitiveness and the creation of an enabling environment that is conducive to private-sector investment in research, development and innovation for the upgrading of food fermentation processes to respond to market demand.
- Establishment and strengthening of the research and infrastructural support base for work on starter culture development, bioreactor design and for the development of diagnostic equipment for monitoring food safety and traceability. This infrastructural support base would include laboratories, laboratory equipment, cell bank facilities for the proper preservation and storage of microbial culture preparations.
- Development of scientific data to substantiate the nutritional, health and health-benefit claims associated with fermented foods.
- Establishment of pilot processing facilities for the scale-up and testing of technologies developed in order to facilitate their adoption.

References

S.K. Singh: *Fundamental of Hotel Management and Operations*, Delhi, Centrum Press, 2010.

S.K. Singh: *Strategic Management in Hotel and Restaurant Industry*, Delhi, Centrum Press, 2010.

Udai Veer: *Elements of Food Science*, Anmol Publications, Delhi, 2007.

Bibliography

Anuradha Roy.: *Food and Beverage Management*, Aadi Publications, Jaipur, 2010.

Archana Ruhela.: *Agriculture and Food Security*, Oxford Book Company, Jaipur, 2008.

Arindam Ramaswamy.: *Elements of Food Science*, Oxford Book Company, Jaipur, 2010.

Arindam Ramaswamy.: *Food Science and Quality Control*, Book Enclave, Jaipur, 2008.

Arindam Ramaswamy: *Elements of Food Science*, Oxford Book Company, Jaipur, 2010.

Arnold C. Long.: *Advances in Seafood Biochemistry: Composition and Quality*, Cyber Tech Publications, Delhi, 2009.

Arthur W. Dana.: *Kitchen Planning for Quality Food Service*, Harper Publications, New York, 1949.

Arti Sankhla, Renu Mogra and Prachi Avinash: *A Practical Guide to Food Preservation*, Agrotech Publishing Academy, Jaipur, 2014.

Ashim Gupta.: *Food and Beverage Service Management*, Centrum Press, Delhi, 2012.

Aswini Kartik.: *Food and Beverage Laws*, Alfa Publications, Delhi, 2009.

Aysha Aziz Faridi: *Dairy Microbiology*, Random Publications, Delhi, 2012.

B. K. Mishra: *Dairy and Food Processing Industry : Recent Trends (2 Vols-Set)*, Biotech Books, New Delhi, 2014.

B.D. Sharma.: *Agriculture Development and Food Security*, Ancient Publishing House, Delhi, 2012.

B.K.P. Singh: *Climate Change and Food Safety*, Surendra Publications, Delhi, 2011.

Bernard Davis, Andrew Lockwood.: *Food and Beverage Management*, Sally Stone Publications, 1998.

Bhavana Sabarwal: *Dictionary of Food and Nutrition*, Crescent Publications, Delhi, 2012.

Bhupendra Singh Khatkar.: *Food Science and Technology*, Daya Publications, Delhi, 2007.

Bimal Kumar Kapoor.: *Food Science and Nutrition*, Mohit Publications, Delhi, 2007.

Dev Raj.: *Food Science and Technology: Glossary of Preeminence*, New India Publications, Delhi, 2011.

Dilip Shah: *Co-Operativization Liberalization and Dairy Industry in India*, ABD Publishers, Delhi, 2000.

Doreen Virtue: *Constant Craving: What Your Food Cravings Mean And How To Overcome Them*, Hay House India, 2011.

Ernest R. Vieira.: *Elementary Food Science*, Chapman and Hall, 2010.

Eugene Lyman Fisk, Adelle Davis, Florence Daniel, Ruth A. Wardall and Harry Snyder.: *Food Science and Nutrition (Vols* 1 *to* 2 *Set)*, Shree Publications, Delhi, 2009.

Fraser, Evan D. G.,Rimas, Andrew: *Empires of Food*, Random House, Delhi, 2011.

Fred Searing.: *Profit Building Ideas for Food and Food Service,* Dahl Publishing. Co., Stamford Conn, 1947.

G Chengappa, N Nagaraj and Ramesh Kanwar: *Challenges to Sustainable Agri-Food Systems,* I.K. International Publications, Delhi, 2007.

Gaurav Gandhi.: *Hotel Management Food and Food Services*, Random Publications, Delhi, 2012.

Geeta Budhiraja and Arun Budhiraja: *Bhog: Temple Food of India*, Krishnaprerna Publications, 2012.

Gisbert Otterstatter: *Coloring of Food, Drugs and Cosmetics*, Taylor and Francis Publications, 2010.

H.A. Modi: *Fermented Milk Products*, Aavishkar Publication, Delhi, 2011.

Harmeet Singh: *Dairy Farming*, APH Publication, Delhi, 2011.

Hema Thapar: *Food Science and Health*, Pacific Publications, Delhi, 2011.

Hema Thapar: *Nutrition and Food Science*, Pacific Books International, Delhi, 2011.

J. O. Dahl: *200 Ways to Control Food Costs in Quantity Cookery,* The Dahls, Stamford Conn, 1941.

Jagmohan Negi and Gaurav Manoher: *Food and Beverage Management*, Himalaya Publications, Delhi, 2011.

Jagmohan Negi, M.J. Gaurav and Ritushka and Suniti: *Food And Beverage Service Skills Techniques Operation And Management*, Kanishka Publications, Delhi, 2011.

Jagmohan Negi: *Food and Beverage Law: Food Safety and Hygiene*, Aman Publications, Delhi, 2004.

Jerry D Souza and Jatin Pradhan: *Handbook of Food Science Catering Technology and Kitchen Management*, SBS Publications, Delhi, 2010.

M Lakshmi Narasaiah: *Economic Growth and Food Security*, Discovery Publishing House, Delhi, 2008.

Madhulika Bhatnagar: *Encyclopaedia of Catering Technology, Food Service and Hospitality Management, Vol. I to III*, Anmol Publications, Delhi, 2008.

Manju Malhi: *Easy Indian Cookbook: The Step-By-Step Guide to Deliciously Easy Indian Food at Home*, Viva Books, Delhi, 2008.

Margaret M. Walsh: *Food and Nutrition Manual for Institutions*, Welfare Federation of Cleveland, Cleveland, 1950.

Mridula Mishra: *Agriculture and Food Economics*, Serials Publications, Delhi, 2010.

Parminder K. Bhandari: *Achieving Nutritional Goals*, Sonali Publications, Delhi, 2010.

R.P. Saxena: *Food Services and Catering Management*, Centrum Press, Delhi, 2010.

Ravee Chauhan: *Advanced Hotel Industry and Tourism*, Vista International Publishing House, 2011.

Renu Singh: *Food and Beverage Management*, Delhi, Rajat Publications, 2012.

Roberta Angelone: *Cooking Italian Food in India*, Westland Publisher, 2013.

Roshan Chaudhary: *Basics of Food Chemistry*, Anmol Publications, Delhi, 2008.

S.K. Singh: *Fundamental of Hotel Management and Operations*, Delhi, Centrum Press, 2010.

S.K. Singh: *Strategic Management in Hotel and Restaurant Industry*, Delhi, Centrum Press, 2010.

Udai Veer: *Elements of Food Science*, Anmol Publications, Delhi, 2007.

Index